Statistics 250
Lab Workbook

Fall 2014 - Winter 2015

Weekly Labs, In-Lab Projects,
Supplements,
and Old Exams for Review
Used in all lab sections of Stat 250

Dr. Brenda Gunderson
Department of Statistics
University of Michigan

Michigan Publishing
University of Michigan Library
Ann Arbor

ISBN 978-1-60785-439-5

Table of Contents

Note to Students

Welcome to Statistics 250 at the University of Michigan! This lab workbook is designed for you to use in lab and as extra preparation for exams. In the workbook, you will find the following materials:

Supplemental Material – great summaries for reference throughout the term:

1. SPSS Commands Reference
2. Notation Sheet
3. Name That Scenario
4. Editing Charts in SPSS
5. Important Notes for Hypothesis Testing
6. Interpretation Examples
7. Summary of T-tests and Name That Scenario Practice for Means
8. Regression Output in SPSS

Weekly Labs (numbered 1 to 13) – each lab contains the follow parts:

- **Lab Background** – objective and brief overview material, which is good to take a couple minutes to read *before* you come to lab each week.
- **Warm-Up Activity** – quick questions for you to do before the In-Lab Project, usually a quick review of concepts you have seen in lecture.
- **ILP (In-Lab Project)** – one or more activities you will work on in lab, in groups.
- **Cool-Down Activity** –questions for you to do after the ILP for further reflection and application of the concepts covered in the ILP.
- **Example Exam Questions** – old exam questions on the lab topic for practice.

Note: each week some part of the weekly lab will be your **Ticket out the Door**, collected by your GSI and part of your lab grade.

Old Exams – complete sets of actual old exam for studying. Be sure to refer to CTools to see if any problems on these old exams are not relevant for your particular upcoming exam (due to differences in the semester schedule). This information, in addition to solutions, will be posted on CTools in the "Review Info" folder under the "Resources" tab closer to each exam date.

The Labs are designed to be interactive and to provide you with a complete example for each concept. Completing the corresponding PreLab assignment (a link to video instructions for PreLabs will be on CTools and the Stat 250 YouTube channel) and reading the upcoming lab background overview before lab each week is a good way to prepare for the various lab activities.

Good luck in Statistics 250!
-- *The Stat 250 Instructors and GSIs*

Supplement 1: SPSS Commands Summary
By Lab – For Quick Reference

Lab 1 – Bar Charts, Histograms, Numerical Summaries

Open a data file after having SPSS already open: File> Open> Data

To produce a **Histogram:** Graphs> Legacy Dialogs> Histogram

To generate **Descriptive Statistics I:** Analyze> Descriptive Statistics> Descriptives

To generate **Descriptive Statistics II:** Analyze> Descriptive Statistics> Frequencies

To produce a **Bar Chart:** Graphs> Legacy Dialogs> Bar> Simple> Summaries of separate variables

Lab 2 – Boxplots, Time Plots, Q-Q Plots

To produce a **Boxplot** for a single variable with no groups: Graphs> Legacy Dialogs> Boxplot> Simple> Summaries for separate variables

To use **Data Label Mode:** From inside the Chart Editor, Elements> Data Label Mode

To **Split (or unsplit) the data (get charts and statistics by group):** Data> Split File

To produce **Side-by-Side Boxplots:** Graphs> Legacy Dialogs> Boxplot> Simple> Summaries for groups of cases

To produce a **Sequence (Time) Plot:** Analyze> Forecasting> Sequence Charts

To produce a **Q-Q Plot:** Analyze> Descriptive Statistics> Q-Q Plots

Lab 7 – One-Sample t Procedures for a Population Mean

To produce a **Confidence Interval for a population mean (method I):** Analyze> Descriptive Statistics> Explore> Statistics option

To produce a **Confidence Interval for a population mean (method II):** Analyze> Compare Means> One-Sample T Test

To perform a **One-Sample T Test for a population mean:** Analyze> Compare Means> One-Sample T Test

Lab 8 – Paired t Procedures

To calculate a **confidence interval for μ_D:** Analyze> Compare Means> Paired-Samples T Test

To perform a **Paired T Test:** Analyze> Compare Means> Paired-Samples T Test

To **compute Differences:** Transform> Compute

Lab 9 – Independent Samples t Procedures

To construct a **confidence interval for μ_1 - μ_2:**
Analyze> Compare Means> Independent-Samples T Test

To perform a **Two-Samples T Test**: Analyze> Compare Means> Independent-Samples T Test

Lab 10 – One-way Analysis of Variance (ANOVA)

To perform an **ANOVA:** Analyze> Compare Means> One-Way ANOVA

Lab 11 – Simple Linear Regression

To produce a **Scatterplot:** Graphs> Legacy Dialogs> Scatter/Dot

To perform a **Linear Regression:** Analyze> Regression> Linear

To produce a **Residual plot:** Graphs> > Legacy Dialogs> Scatter/Dot

Lab 12 – Chi-Square Tests

To **weight cases by Counts:** Data> Weight Cases

To perform a **Goodness of Fit Test:** Analyze> Nonparametric Tests> Chi-Square

To perform a **Test of Independence:** Analyze> Descriptive Statistics> Crosstabs

To perform a **Test of Homogeneity:** Analyze> Descriptive Statistics> Crosstabs

Supplement 2: Notation Sheet

The table below defines important notations, including that used by SPSS, which you will come across in the course. This is not an exhaustive list, but it is a fairly comprehensive overview of the "strange letters" used in the course.
Note: Blank cells mean there is no corresponding notation.

Name	Population Notation	Sample Notation	Notation used in SPSS
Summary Measures			
Mean	μ (read as "mu")	$\bar{x}$ (*x*-bar)	Mean
Proportion	p	$\hat{p}$ (*p*-hat)	
Standard deviation	σ (sigma)	s	Std. Deviation
Variance	σ^2	s^2	Variance
Sample size		n	N
Confidence Intervals			
Multipliers		z^* (*z*-star)	
		t^* (*t*-star)	
Margin of error		m, m.e.	
Hypothesis Testing			
Test statistics		z	
Note: *t*, *F*, and χ^2 statistics have		t	t
degrees of freedom		F	F
(abbreviated df) associated with them. Look for these on your Formula Card.		χ^2 (chi-square)	Chi-square
Significance level		α (alpha)	
***p*-value**		*p*-value	Sig.

Name	Population Notation	Sample Notation	Notation used in SPSS
Analysis of Variance (abbreviated ANOVA)			
Sum of squares for groups		SSG	Between groups (look in the column labeled Sum of Squares)
Sum of squares for error		SSE	Within groups (look in the column labeled Sum of Squares)
Mean square for groups		MSG	Between groups (look in the column labeled Mean Square)
Mean square error		MSE	Within groups (look in the column labeled Mean Square)
Regression			
Response (dependent) variable	y	y	(given by name of y-variable)
Predicted (estimated) response	E(y) (expected value of y)	$\hat{y}$ (y-hat)	
Explanatory (independent) variable	x	x	(given by name of x-variable)
y-intercept	β_0 (beta-not)	b_0	B (look in the row labeled (Constant))
Slope	β_1 (beta-one)	b_1	B (look in the row labeled with the name of the x-variable)
Coefficient of correlation		r	R, Beta
Coefficient of determination		r^2	R Square
Error terms vs Residuals	ε (error terms)	e (residuals)	Unstandardized residuals

Supplement 3: Name That Scenario

The first thing to do in any research inference problem is determine what type of inference problem it is. This will help in deciding what procedure/formulas are appropriate to use. The following questions can help you determine the data scenario you are working with.

Please note, when answering, "How many variables are there?" do not count the variable which defines the populations (if there is more than one population).

❑ How many populations are there?

One **Two** **More than two**

❑ How many variables are there?

One **Two**

❑ What type of variable(s)?

Categorical **Quantitative**

❑ Then use the following table to determine which type of inference would be appropriate for this scenario.

Note the corresponding parameter is in parentheses, where appropriate.

Number of Variables and Type		Number of Populations: One	Two	More Than Two
One	**Categorical**	❑ 1-sample inference for population proportion (p) (Labs 5 an 6) ❑ Chi-square: Goodness of Fit (Lab 13)	❑ 2 indep. samples inference for the difference between 2 population proportions ($p_1 - p_2$) ❑ Chi-square: Homogeneity (Lab 13)	❑ Chi-square: Homogeneity (Lab 13)
	Quantitative	❑ 1-sample inference for population mean (μ) (Lab 8) ❑ Paired samples inference for a population mean difference (μ_D) (Lab 9)	❑ 2 indep. samples inference for the difference between 2 population means (μ_1 - μ_2) (Lab 10)	❑ ANOVA (μ_i – where there is one μ_i for each population) (Lab 11)
Two	**Categorical** (relationship)	❑ Chi-square: Independence (Lab 13)		
	Quantitative (relationship)	❑ Regression (β_1) (Lab 12)		

Supplement 4: Editing Charts in SPSS

Once we have a histogram (or any chart) made, we may wish to edit the chart (perhaps to change the color of the bars or change the number of class intervals). To do this, double click on the chart displayed in the output viewer window. This will open the chart in the SPSS **Chart Editor window**.

Suppose we want to **change the color** of the histogram bars from tan to a light green. To do this, double click on one of the tan bars and click on the **Fill & Border tab** in the properties window. Change the **fill** color to light green (click on the light green box color) and click on **apply** and then **close**.

To **change some aspect of the *x*-axis**, such as the scaling, double click on the *x*-axis and its corresponding properties box will appear with many tab options. The **scale tab** can be used to change the endpoints and major increments of the *x*-axis value labels. You could adjust the minimum to 10,000 and the maximum to 150,000. Leaving the Auto box checked for the Major Increment option will let SPSS create the increment size. The **Histogram Options tab** would allow you to add a normal curve to a histogram, as well as change the starting position and size of the bins (classes or intervals) in the histogram. In general, SPSS uses algorithms to produce a nice display of the data. These options are helpful if you have multiple plots that you would like to display using the same x-axis values so comparisons can be more easily made. You can also change the gray background color to white under the **Properties window** with the background highlighted. The **help** button in the lower right corner of the Properties box can be selected to provide more details about any of the various options for that tab. Once you have finished customizing your chart, you can close out the chart editor.

There are alternate ways to get to the Properties box in order to customize your chart. Once you have double-clicked on your chart to open the Chart Editor, click once on the part of the chart that you wish to customize (so that is it highlighted). Then, click on the **Show Properties Window** tab (it looks like a paint palette) in your menu. This will open the Properties box. Another alternative is to simply select **Properties** under the **Edit** menu in the Chart Editor. Also, note that if you do not close the Properties box, and you continue highlighting different parts of your chart, the Properties box updates so that you can customize those parts as you go.

For **boxplots**, if there are any points denoted as outliers, you can identify them by looking at their case label number in the default output. The Chart Editor provides a special mode for identifying individual cases whose data labels you want to display. This is the **data label mode**, and when you are in data label mode, you can't change anything else in the chart. From the menus, choose **Elements> Data Label Mode**. The cursor changes shape to indicate that you are in data label mode. Click the data element for which you want to display the case label. If there are overlapping data elements in the spot that you click, the Chart Editor displays the **Select Data Element to Label** dialog box. This dialog allows you to select the specific data element or elements for which you want to display data labels. The Chart Editor displays the data label in a default position related to the data element. When you are finished choosing data elements, from the menus choose **Elements> Data Label Mode** again, and the cursor changes back to the arrow to indicate that you are no longer in Data Label Mode.

The **Options** menu lets you customize your chart further. You may add a title or text box from this menu. Text boxes can appear anywhere in a chart. From the Chart Editor menus, select **Options> Text Box** or **Options> Title** depending on which you want. For titles, the Chart Editor creates the title box and automatically positions it in the top center of the chart. Type the text and press enter when you are finished typing. To enter line breaks, press Shift+Enter. If necessary, use the **Text** tab to format the text. For text boxes, you can drag and drop to reposition them. You may need to resize the graph so the text box will not cover up part of the graph. You can also copy the plots onto MSWord or another text editor and then type in your name and title within the document.

Saving Output Boxes and Graphs

Images and other output from SPSS can often be copied and then pasted into a document by selecting the desired output, right-clicking, and choosing **Copy**.

1. ***To save an output box***, such as a table of descriptive statistics, first have the location where you would like to store the output open. Then right-click on the output table and select **Copy**. The table can then be pasted into a document or text-field (such as those in your PreLab assignments or homework). If you are pasting into a Word document and if the output does not appear to format correctly, it may be a good idea to choose **Paste Special** and paste as an image. Your stats 250 GSIs prefer not to receive Word documents.
2. ***To save a graph***, such as a boxplot or histogram, you will want to **Export** the graph as an image. Select the graph you wish to export then select **Export** from the **File** menu. At the top click the **Selection** button, select **None (graphics only)** under **Document**, and then choose the file type at the bottom. For uploading to www.lecturebook.com, the extension ".jpg" or ".png" is required. Before completing the export command, be sure to give your file an informative name and note the location where the file will be saved.
3. ***To save an entire SPSS session***, you can export output (all output in the viewer, charts only or text only) in many possible formats (html, jpeg, bitmap, etc.). You first make the Viewer the active window and select the **Export** command is under the **File** menu.

You can also **print** the contents of your output viewer window (all output, text as well as charts) or any selected portion. Click on **File** from the menu bar and then choose **Print**. You could also just save your output file within SPSS by selecting the **Save as** option and giving a name for your file and click **OK**.

Supplement 5: Notes about SPSS *t* Procedures

1. The reported *p*-value under the column heading of ***Sig. (2-tailed)*** is for a 2-sided test. For a one-sided test, you first divide the reported 2-tailed *p*-value in half ($p/2$). If the *t*-statistic is positive and the alternative hypothesis was upper-tailed (>), then $p/2$ is the *p*-value. If the *t*-statistic is negative and the alternative hypothesis was lower-tailed (<), then $p/2$ is the *p*-value. However, if the *t*-statistic is positive and the alternative hypothesis was lower-tailed (<) or if the *t*-statistic is negative and the alternative hypothesis was upper-tailed (>), then the *p*-value is $1 - p/2$.

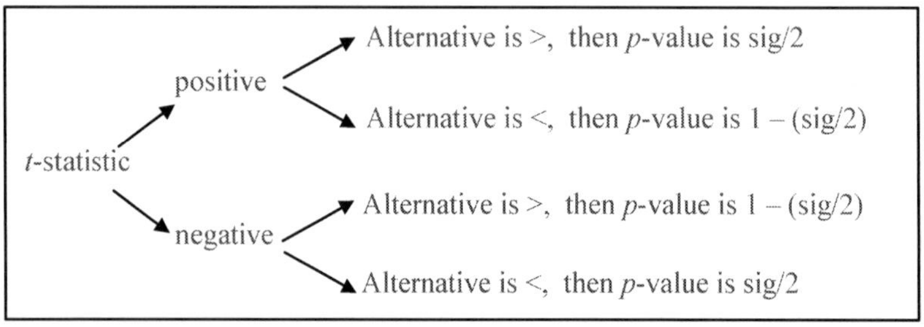

For example, consider a 2-sided test with an observed *t* statistic value of 0.948 and a *p*-value of 0.364. This 0.364 is actually the sum of two equal areas: one being the area to the right of 0.948 and the other being the area to the left of -0.948 under the *t*-distribution with df = 11 curve (see Figure 1 below).

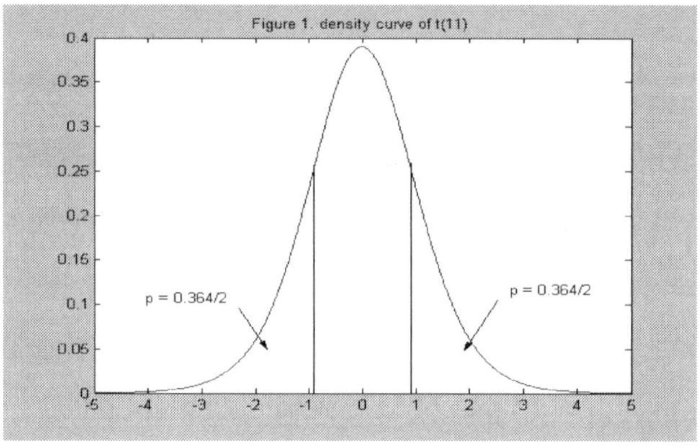

If the alternative hypothesis had been upper-tailed (>), then the *p*-value would be only the area to the right (in the direction of extreme) of the observed *t*-statistic of 0.948, which is half of the two-sided *p*-value, or 0.182. But if the alternative hypothesis had been lower-tailed (<), the *p*-value would be the area to the left (in the direction of extreme) of the observed *t*-statistic of 0.948. So the *p*-value would be $1 - (0.364/2) = 0.818$.

2. Sometimes SPSS will display a *p*-value of 0.000. Clearly, the probability is not exactly zero. Rather, it is zero to 3 significant digits. Thus, it is correct to say that the *p*-value is less than 0.0005, since anything greater when rounded would have resulted in **Sig. 2-tailed *p*-value** of 0.001 or more.

3. One can use a one-sample t $(1 - \alpha)\%$ confidence interval to test a two-sided hypothesis at the α % level by checking whether the hypothesized or null test value is contained in the interval. If it is, we cannot reject H_0, but if not, we would reject H_0. Recall that if you want to produce a confidence interval for the population mean, μ, you **must** specify the test value to be 0 in the one-sample *t*-test dialog box. For example, a test value of 3.0 that would result in a 95% confidence interval of the difference given as (-0.265, 0.665) is actually a confidence interval for $\mu - 3.0$. See Lab 5 for more details.

4. Supplement 6 provides an overview of the assumptions for all three *t*-tests that are presented in Labs 8, 9, and 10.

Supplement 6: Interpretation Examples

In 1980, Bausch and Lomb Corporation developed a new type of extended-life contact lens made of silicone, which it claimed had a useful life of more than 4 years. During the research and development period, a random sample of 6 contact wearers was asked to wear the new contact lenses and record how long they lasted. The average useful life of the six pairs of lenses was 4.6 years, with a standard deviation of 0.49 years.

a. **Interpretation of the Standard Deviation *s*:**
The **average distance** of the observed useful lives of these lenses **from their mean** useful life of 4.6 years is ***about*** 0.49 years.

b. Calculate the value of the **standard error of the mean.**

$$SE(\overline{X}) = \frac{s}{\sqrt{n}} = \frac{0.49}{\sqrt{6}} = 0.200$$

Interpretation:
We would **estimate** that the **average distance** of the possible sample mean useful life values (obtained from repeated random samples of size $n = 6$ pairs of such lenses) from the **population mean useful life,** μ, to be ***roughly*** 0.20 years.

c. Construct a **90% confidence interval** for the population mean life of all such silicone-based lenses:

$$4.6 \pm (2.015)(0.200) \Rightarrow (4.197, 5.003)$$

Interpretation of the Interval:
This interval provides a range of reasonable values for the population mean useful life, μ. We would estimate the **population mean** useful life, μ, to be between 4.197 years and 5.003 years, with 90% confidence.

Interpretation of the 90% Confidence Level:
This interval was made with a method, which if repeated, would generate many 90% confidence intervals. We would expect 90% of these resulting intervals to contain the population mean life, μ.

d. State the hypotheses to test the claim made by Bausch and Lomb about their new contact lens; that is, test if the population mean useful life is more than 4 years.

$H_0 : \mu = 4$, $H_a : \mu > 4$, with an observed t-test statistic of

$$t = \frac{\bar{x} - \mu_0}{s/\sqrt{n}} = \frac{4.6 - 4}{0.200} \cong 3.00 .$$

The ***p*-value** for this test is the probability of getting a *t*-test statistic at least as extreme as the observed test statistic, assuming the null hypothesis is true. So we have the *p*-value is $P(T \geq 3.00)$ found under the *t*(5) distribution. This *p*-value turns out to be equal to 0.015.

Interpretation of the value of the test statistic *t = 3.00* in terms of a distance: The observed sample mean was 3 average distances (i.e. 3 standard errors) above the hypothesized mean of 4.

Interpretation of the resulting *p-value of 0.015*: If the null hypothesis was true (the population mean useful life is just 4 years) and this procedure (study) was repeated many times, we would expect to see a *t*-test statistic value of 3.00 or larger in only 1.5% of the repetitions. Thus are data are somewhat unusual under the null hypothesis theory, providing evidence for the alternative theory that the population mean useful life is greater than 4 years.

e. At a 10% significance level, what is the **decision**?
Reject H_0 since the *p*-value is less than 0.10.

f. What is the **conclusion**? There is sufficient evidence to conclude that the **population mean** useful life of the new lenses is greater than 4 years.

> ***NOTE:***
> These interpretations can be extended to the any test and confidence interval, adjusting for the different parameters, different directions of extreme, different test statistics, etc.

Supplement 7: Summary of the Main *t*-Tests

The three inference scenarios presented in Labs 8, 9, and 10 are: **one-sample *t* procedures, paired *t* procedures, and two independent samples *t* procedures**. It is important to look at the data to determine if doing a particular *t* procedure is appropriate. That is, we need to check the assumptions. (Recall that checking assumptions is the second step in performing a hypothesis test.)

The *t* procedures have the following general assumptions:

1. Each sample is a **random sample** – (the observations can be viewed as realizations of independent and identically distributed random variables). In the paired *t* procedures, the **differences** are assumed a random sample.

2. Each sample is drawn from a **normal population**, that is, the response variable has a normal distribution for each population. In the paired *t* procedures, the population of **differences** is assumed to have a normal distribution. In the two-sample case, both populations of responses are assumed to have normal distributions.

   You need normality of the underlying population for the response in order to have normality for the sample mean. In the case where you do not have a normal population, you can still have normality of the sample mean if you have a large enough sample size (most texts state that a sample size of *at least 30* is required). In Lab 7 (Sampling Distributions and the CLT) you work with an applet that demonstrates the CLT using a *large* sample size of 25. Thus we will accept *at least 25* as large enough for inference about means.

3. For the two independent samples *t* procedures, we also assume that the **two samples are independent**. We also need to assess whether the **two population variances** can be assumed **equal** in order to decide between the pooled and the unpooled *t* tests.

Graphical tools can be used to check these assumptions (see Labs 2 and 3 for more details about these various graphs).

Time Plots (or Sequence Plots): If your quantitative data have been gathered over time, then a time plot can be used to determine if the underlying process that generated that time dependent data ***appears*** to be stable. So these plots can help us assess if the random sample assumption is reasonable. For the paired design problems, we assume our set of differences calculated from the paired observations ($D_1, D_2, \ldots, D_n$) are a random sample. So if these differences were obtained over time, they should be plotted against their order to see if they look like they came from one population of all differences (no changing mean or variability over time).

Remember:

#1 Time or Sequence plots are useful for checking stability only when the data are ordered in some sense. If there is no inherent order to the data, a sequence plot should not be made.

#2 If a Time plot makes sense to be examined and does show evidence of instability, it would not make sense to treat those observations as being a random sample; thus it would not be appropriate to make a histogram, Q-Q plot, or boxplot of the observations.

Histograms: Histograms are especially useful for displaying the distribution of a quantitative response variable. You could make a histogram of the observations in a one-sample problem, of the differences in a matched pairs design, and of each of the two samples separately in the independent samples design. Examine the histogram for evidence of strong departures from normality, such as bimodality or extreme outliers. Since you are just plotting data (just a sample and not the entire population of responses), your histogram may not look *perfectly* bell-shaped or normal.

Q-Q plots: Q-Q plots (or quantile plots or normal probability plots) are generally better than histograms for assessing if a normal model is appropriate. If the points in a Q-Q plot fall *approximately* in a straight line (with a positive slope) then the normal model assumption is reasonable.

Boxplots: Boxplots are most useful for assessing the validity of the assumption of **equality of population variances** in **the two independent samples design**. We would see if the IQRs (shown graphically by the length of the boxes) are comparable, and also compare the overall ranges. If they do have comparable lengths or sizes (they do not need to be lined up), then we have support that the equality of population variances assumption is reasonable. We would also want to compare the two sample standard deviations themselves, and Levene's test of equality of the two population variances may also be available.

Name that Scenario Practice for the Three T Tests:

Having just reviewed the three main *t*-test inference scenarios, you should understand the testing procedures and be able to interpret the results of a test. However, it is important to know when each scenario applies. Read each of the following inference scenarios and determine which of the three *t*-test procedures would be most appropriate: the one-sample *t*-test, the paired *t*-test, or the two-independent samples *t*-test.

1. A researcher is studying the effect of a new teaching technique for middle school students. One class of 30 students is taught using the new technique and their mean score on a standardized test is compared to the mean score of another class of 27 students who were taught using the old technique.

2. A company claims that the economy size version of their product contains 32 ounces. A consumer group decides to test the claim by examining a random sample of 100 economy size boxes of the product, since they have received reports that the boxes contain less than the 32 ounces claimed.

3. At some universities, athletic departments have come under fire for low academic achievement among their athletes. An athletic director decides to test whether or not athletes do in fact have lower GPAs. A random sample of 200 student athletes and a random sample of 500 non-athlete students are taken and their GPAs are recorded.

4. As part of a biology project, some high school students compare heart rates of 40 of their classmates before and after running a mile. They want to see if the heart rate of students their age is faster after running a mile than before, on average.

5. A hospital is studying patient costs; they decide to follow 500 surgery patients' hospital and medical bills for a year after surgery, and compare them to the estimated costs provided to the patients before surgery. They want to see if the estimated and actual costs are comparable on average.

6. A chemical process requires that no more than 23 grams of an ingredient be added to a batch before the first hour of the process is complete. An analyst feels that due to current settings more than 23 grams may actually be added. If the analyst is correct, the settings need to be altered and recent batches recalled. A random sample of 25 batches is obtained from the machine that is supposed to add the ingredient. The measurements are used to test the analyst's claim.

Supplement 8: Regression Output in SPSS

There are four parts to the default regression output. Use the scroll bar at the right edge of the **Output Window** to scroll up to the top of the regression output. The first section just reminds you which variable was entered as the explanatory x variable; for this example, the explanatory variable is ***DNA***.

The second section has the heading **Model Summary**. The Model Summary starts with the correlation between the two variables, **R**, which is the ***absolute value*** of the **correlation coefficient**, *r*. You need to look at the **sign of the slope** of the regression line to determine if you need to put a minus sign in front of this value to correctly report the correlation coefficient. (The actual value of the correlation coefficient is also reported in the last section of regression output, under the column heading **Beta**.) The correlation coefficient measures the strength of the linear association between the two variables. The closer it is to +1 or -1, the stronger the linear association. The square of the correlation, the **R Square** quantity, has a useful interpretation in regression. It is often called the **coefficient of determination** and measures the proportion of the variation in the response that can be explained by the linear regression of *y* on *x*. Thus, it is a measure of how well the linear regression model fits the data. The **Std. Error of the Estimate** gives the value of *s*, the estimate of the population standard deviation σ.

Model Summary

Model	R	R Square	Adjusted R Square	Std. Error of the Estimate
1	.856 [a]	.732	.699	4.851

a. Predictors: (Constant), DNA

The third part of the output contains the **ANOVA table for regression**, used for assessing if the slope is significantly different from 0 via an *F* test. The corresponding *t*-test will be discussed first and we return to this ANOVA part later.

ANOVA [b]

Model		Sum of Squares	df	Mean Square	F	Sig.
1	Regression	515.141	1	515.141	21.894	.002 [a]
	Residual	188.228	8	23.528		
	Total	703.369	9			

a. Predictors: (Constant), DNA

b. Dependent Variable: PLAQUE

The last portion of the output falls under the heading **Coefficients**. In this section, the least square estimates for the regression line are given. These estimated regression coefficients are found under the column labeled **B**. The estimated slope is next to the independent variable name (in this example it is DNA), and the estimated intercept is next to **(Constant)**. So, b_0 is the coefficient for the variable (Constant), and b_1 is the coefficient for the independent variable x in the model. The next column heading is **Std. Error**, which provides the corresponding **standard error** of each of the least squares estimates. Also produced in this table, are the ***t*-test statistics** in the column labeled **t** and **Sig.**, which reports the two-sided ***p*-values** for these *t*-test statistics.

Coefficients [a]

		Unstandardized Coefficients		Standardized Coefficients		
Model		B	Std. Error	Beta	t	Sig.
1	(Constant)	-.548	8.193		-.067	.948
	DNA	.167	.036	.856	4.679	.002

a. Dependent Variable: PLAQUE

The ***t*-statistic** for the slope, in the second row, is a test of the significance of the model with ***x*** versus the model without ***x***, that is, for testing H_0: $\beta_1 = 0$ versus H_a: $\beta_1 \neq 0$. The ***t*-statistic** for the *y*-intercept, in the first row, is a test of whether the *y*-intercept (β_0) is different from zero. This test is not often of interest unless a value of 0 for the *y*-intercept is meaningful and of interest. For example, if x = amount of soap used and y = height of the suds, then an intercept value of 0 is meaningful as no soap would lead to no suds. The column labeled **Sig.** gives the **two-sided *p*-value** for the corresponding hypothesis test.

SPSS also provides the information to calculate confidence intervals for the parameter estimates. The column labeled **Std. Error** provides **standard errors** (estimated standard deviations) of the parameter estimates and is the quantity that is multiplied by the appropriate ***t** value** in computing the half-width of the confidence interval. Recall that you can request SPSS to produce these confidence intervals for you using the **Statistics** button in the **Regression dialog box**.

Interpretation of estimated slope b_1:
According to our regression model, we estimate that increasing DNA by one unit has the effect of increasing the predicted plaque by .167 units.

Interpretation of r^2:
According to our model, **73% of variation** in plaque levels can be accounted for by its **linear relationship** with DNA.

Decision for test of a significant linear relationship:
Since the *p*-value = .002 is less than the significance level $\alpha = .05$, we can reject the null hypothesis that the population slope, β_1, equals 0.

Conclusion: There is sufficient evidence to conclude that in the linear model for plaque based on DNA the population slope, β_1, does not equal zero. Hence, it appears that DNA is a significant linear predictor of plaque.

Let's return to the **ANOVA table** in the middle of the regression output.

ANOVA [b]

Model		Sum of Squares	df	Mean Square	F	Sig.
1	Regression	515.141	1	515.141	21.894	.002 [a]
	Residual	188.228	8	23.528		
	Total	703.369	9			

a. Predictors: (Constant), DNA

b. Dependent Variable: PLAQUE

The **Regression Sum of Squares** corresponds to the portion of the total variation in the data that is accounted for by the regression line. Everything that is left over and not accounted for by the regression line is placed in the **Residual Sum of Squares** category. Then, dividing the sum of squares by their respective **df** (degrees of freedom) yields the **Mean Squares**.

Finally, the ratio of the Mean Squares provides the ***F*** statistic which tests if the slope is significantly different from zero (i.e. if there is a significant non-zero linear relationship between the two variables – H_0: $\beta_1 = 0$ versus H_a: $\beta_1 \neq 0$.) The **Sig.** is the corresponding *p*-value for the *F* test of these hypotheses.

In simple linear regression, the *t*-test in the **Coefficients** output for the slope is equivalent to the ANOVA *F*-test. Notice that the square of the *t*-statistic for testing about the slope is equal to the *F*-statistic in the ANOVA table, and the corresponding *p*-values are the same

Checking the Simple Linear Regression Assumptions

Here is a summary of some graphical procedures that are useful in detecting departures from the assumptions underlying the simple linear regression model.

1. **LINEARITY:** Do a scatter plot of *y* versus *x*.
   The plot should appear to be roughly linear.

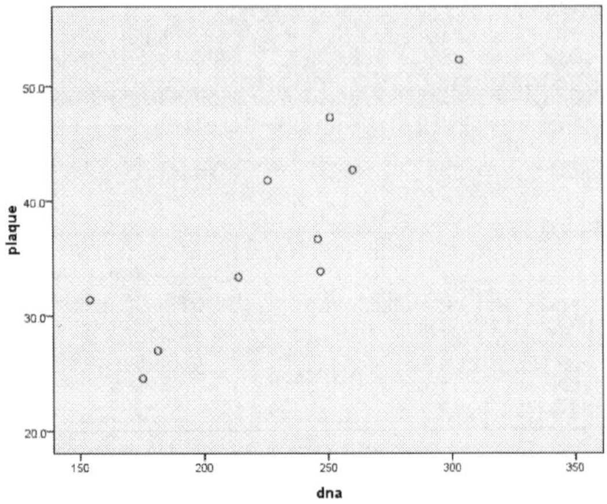

2. **STABILITY:** Do a sequence plot of the residuals.
   The plot should show no pattern indicating any trend in the mean or in the variance of the residuals. An example series plot is shown below. *Remember that it is only appropriate to make sequence plots when there is some ordering present in the data.*

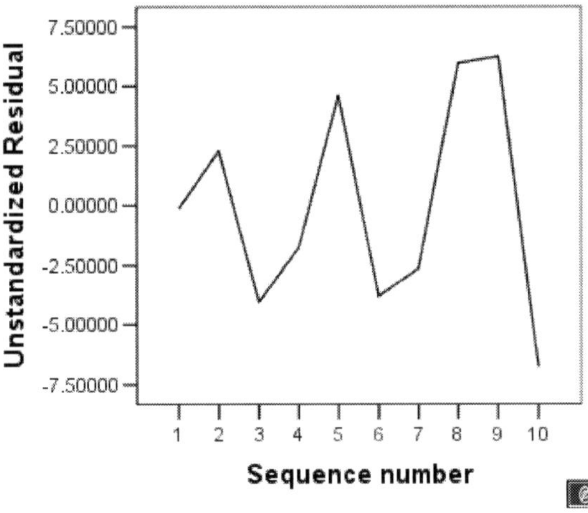

3. **NORMALITY:** Examine a Q-Q plot of the residuals to check on the assumption of normality for the population (true) error terms. An example Q-Q plot is shown below.

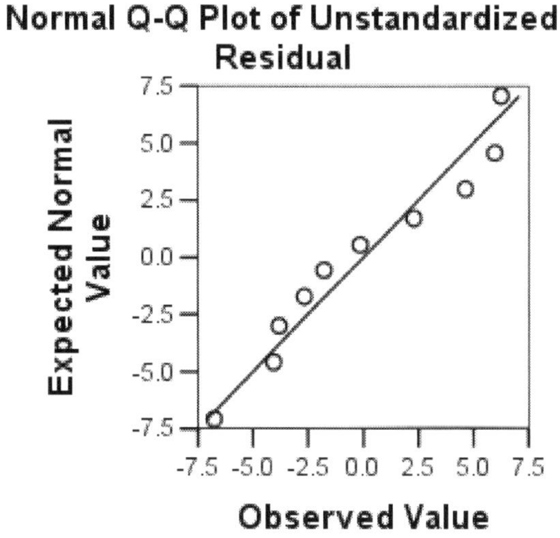

4. **CONSTANT STANDARD DEVIATION** of the population (true) error terms: Make a plot of the residuals versus *x*. This plot is called a **residual plot**. The residuals represent what is left over after the linear model has been fit. The residual plot should be a random scatter of points in roughly a horizontal band, with no apparent pattern. An example residual plot is shown at the right. Sometimes this plot can also reveal departures from linearity (i.e. that the regression analysis is not appropriate due to lack of a linear relationship).

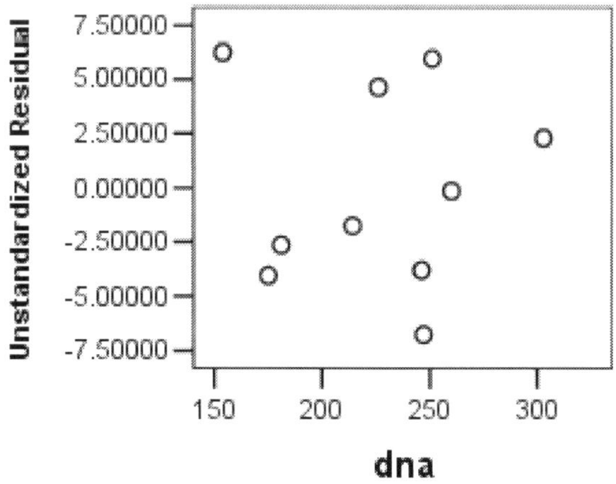

Lab 1: Describing Data with Graphs and Numbers
Histograms, Mean, Median, and Range

Objective: In this course you will be using various forms of technology for exploring and learning statistical concepts. In your first lab you will complete a Scavenger Hunt that will allow you to become familiar with the many class resources. For your In-Lab-Project (ILP) you will use some graphical and numerical tools to summarize the distribution for a quantitative variable or response – a histogram, mean, median, and range. These techniques can be very useful at the start of data analysis to get a feel for the data.

Application: The United States Commerce Department wants to send out a press release about the income growth rate for the nation in 2012. For each state as well as the District of Columbia, they compute the percentage growth of income for the state or district. They have 51 growth rates in their data set and they need to choose a measure of center to include in their report. All the income growth rates range from 2.4% to 5.8% with the exception of one state: North Dakota. This state reported a growth in income of 18.7%. Which measure of center would be more appropriate for the Commerce Department to report given that they have such a large outlier in North Dakota?

Overview: When used properly, graphs can be a very effective way to summarize data. Data on a single quantitative variable should first be examined graphically. One useful graph to summarize the distribution for a single quantitative variable or response is a **histogram**. The overall shape of the distribution and existence of outliers can generally be used to assess if the data appear to be coming from a relatively homogenous population. If so, then various numerical summaries may be used to characterize the center of the distribution (such as mean and median) and the spread of the distribution (such as the range and standard deviation).

Histograms: A histogram displays the distribution of a quantitative variable by showing the frequency (count) or percent of the values that are in various classes. The classes are typically intervals of numbers that cover the full range of the variable. Histograms can be used to assess the **symmetry** and **modality** of a single distribution or for comparing the relative locations and shapes of several distributions.

Measures of Center: Measures of center are numerical values that tend to report the middle of a data set. The two that we will focus on are the mean and the median.

1. **Mean:** The mean of a set of *n* observations is simply the sum of the observations divided by the number of observations, *n*.

2. **Median:** The median of a set of observations, ordered from smallest to largest, is a value such that at least half of the observations are less than or equal to that value and at least half the observations are greater than or equal to that value. The median is less sensitive to outliers than the mean.

Going back to the application example, since the income growth rate of North Dakota is an outlier in the dataset, the mean income growth rate will be influenced by this high growth rate. The mean income growth rate for all 50 states and D.C. is 4.23% while the median income growth rate is 3.8%. The median income growth rate gives the reader a better understanding of the income growth rate across all 50 states and the District of Columbia.

Measures of Variation or Spread: Measures of variation include the range, the IQR and standard deviation. These numerical summaries describe the amount of spread that is found among the data, with larger values indicating more variability. In this lab we will just look at the range, and the next lab will focus on the IQR and standard deviation.

1. **Range:** The range measures the spread of all the data. It is defined as the difference between the maximum value and the minimum value.

2. **IQR:** The IQR measures the spread of the middle 50% of the data. It is defined as the difference between the 3rd quartile (Q3) and the 1st quartile (Q1). These quartiles are also called the 75th and 25th percentiles, respectively. IQR = Q3 – Q1.

3. **Standard Deviation:** Standard deviation is a measure of the spread of the observations from the mean. It is actually the square root of an average of the squared deviations of the observations from the mean. *We can think of the standard deviation as approximately an average distance of the observations from the mean.* The standard deviation can be influenced by outliers.

Warm-Up: Mean and Median

We are interested in analyzing Lebron James' scoring output by game. We have his scoring output for 5 games which we have arranged from lowest to highest: 6, 24, 28, 34, 36. There are two measures of center we could report.

Which measure would be better to report? Median Mean

Which is the most likely value for the mean? 8 12 26

What is the median? 24 28 34

ILP: Visualizing and Exploring Quantitative Data

In this In-Lab Project, you will learn how to create graphs and obtain basic descriptive statistics for quantitative data using SPSS.

Task: The data set **employee data.sav** contains information on employees at a company. You will create an appropriate graph and obtain some descriptive statistics for current salary, and discuss the results.

1. Log onto your computer. To obtain the data set, go to your course site to find the datasets folder. Select **employee data.sav** and save it to a directory of your choice (alternatively you may open the data set directly, in which case you do not need to open SPSS after). If you have saved the data set, go to **Programs**, followed by **Statistics Packages & Math Programs** and then select **SPSS**.

2. To open the **employee data.sav** data set from within SPSS, select the option **Open an existing data source** from the dialog box with the **More Files** line highlighted and click on **OK.** Change the directory to where you saved the data set, select **employee data.sav** and click on the **Open** button. The data set will open, and you can view it (it works like an Excel spreadsheet).

3. The starting view of the data is the **Data Editor** window. Here, you can see the variables in the data set and their values. The first variable you should see is ID.

   What is the second variable present in the data set?

   What type of variable is it (quantitative or categorical)?

   What is the eighth variable present in the data set?

   What type of variable is it (quantitative or categorical)?

4. Create a histogram for current salary. Use the graphs menu - **Graphs> Legacy Dialogs> Histogram**, select (current) salary, and move it to the variable box. Editing details can be found in the **Editing Charts in SPSS** section (Supplement 4) of this workbook.

Note: *All Statistics 250 homework and labwork will require that students provide an appropriate title and their name on each SPSS chart or output. For histograms, click on the* ***Titles*** *button and enter the corresponding information and click on* ***Continue****.*

In the space below make a quick sketch of the resulting histogram and describe what the histogram shows regarding the distribution of current salary. A good description includes information regarding the shape, general center and spread of the values, and if relevant, a comment about possible outliers.

5. Numerical summaries may also be obtained for any quantitative variable. To obtain the five-number summary do **Analyze> Descriptive Statistics> Frequencies** and then choose the summary measures you want under the **Statistics** button.

   Fill in a few of the basic summary measures for current salary.
   Be sure you include appropriate units.

   Mean: ____________________ Median: ____________________

   Min: ____________________ Max: ____________________

   Range: Max – Min = ____________________

Cool-down: Which Measure of Center to Report?

Mark is a Stats 250 GSI who would like to report a measure of center for scores on the first exam. The mean score for his lab section was 77.46 points and the median was 84 points. One of Mark's students did not take the exam and received a zero. Since Mark knows this score will not count against the student, he removes the score of zero from his data.

How will the mean test score change if the grade of 0 is not included?

If there is an outlier test score of 20 points,
which measure of center would you recommend that Mark report?

Example Exam Question on Mean and Median

Tattoos – The Pew Research Center took a survey of Millenials, that is, young adults between the ages of 18 and 29. The survey looked at characteristics that researchers thought described Millenials – social networking, life priorities, and aspects of the respondents' appearances.

One question asked how many tattoos respondents had, and there were 486 respondents who had at least one tattoo. Data for these 486 respondents are shown in the histogram below.

a. How could this histogram be described?
Choose all that apply.

RIGHT SKEWED *LEFT SKEWED*

SYMMETRIC *STABLE*

BIMODAL *UNIMODAL*

DECREASING TREND

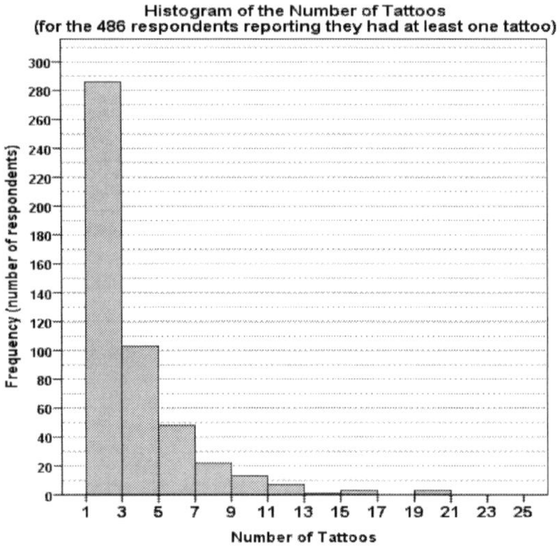

Note: the first interval is [1, 3).

b. If the mean number of tattoos for these Millenials is 3.01 tattoos, which of the following is a reasonable value for the median?

2 *3.01* *4* *5*

c. Is the following statement ***true or false***?

Based on the histogram,
we can be sure the range of the tattoo data is exactly 20 tattoos.

TRUE *FALSE*

Lab 2: Describing Data with Graphs and Numbers
Boxplots, Standard Deviation, QQ Plots, Time Plots

Objective: In this lab, you will use some additional graphical and numerical tools to summarize the distribution for a variable or response. For a quantitative variable, you might make a histogram or boxplot, and report a mean or median value, along with a standard deviation or the interquartile range (IQR). You will be introduced to side-by-side boxplots for comparing two or more distributions for a quantitative response, and bar charts for summarizing categorical data. Additional graphs you might need to examine include time plots for data collected over time and Q-Q plots for checking whether a normal (bell-curve) model is a reasonable distribution for a quantitative variable. These techniques can be very useful at the start of data analysis to get a feel for the data.

Application #1: Jane is a researcher for Northeastern University who wants to learn more about public opinion regarding higher education. She conducts a survey of 1000 adults and asks various questions about higher education. One of the question asked participants to "rate the current job the higher education system in this country is doing in terms of preparing recent college graduates for the workforce". Possible responses to this question were: Excellent, Good, Fair and Poor. Which of the graph should be used to illustrate the results of this particular question?

Application #2: Brad is the manager of the Detroit Tigers and to prepare for the next round of the playoffs, he would like to run a hypothesis test involving the mean number of runs his players have scored over the last month of play. Brad knows that one assumption required for performing this analysis about the mean is that his data must be considered a random sample (the observations can be viewed as coming from the same parent population). He can examine this assumption by collecting the number of runs scored over the last month and creating a time plot.

Overview: For categorical variables, a **bar chart** can be used to display how many fall in each category (frequency distribution). Data on a quantitative variable should also be examined graphically. If the data has been collected over time, the first graph to examine is a **time plot**. If the resulting time plot appears to be stable, or if the data was not collected over time, then two graphs that can be used to summarize the distribution for a single quantitative variable or response are a **histogram,** a **boxplot**, and perhaps a **Q-Q plot**. Each graph provides different information about the distribution. The overall shape of the distribution and existence of outliers can generally be used to assess if the data appear to be coming from a relatively homogenous population. If so, then various **numerical summaries** may be used to characterize the center of the distribution and the spread of the distribution.

Note that some graphical tools are introduced solely in lab, not in lecture, so it will benefit you to read this overview thoroughly

Boxplots: One plot that can detect extreme observations or outliers is the **boxplot.** A boxplot is a graphical representation of the five-number summary, namely, the minimum, first quartile, median (second quartile), third quartile, and maximum of the data. The centerline of the box marks the median (50th percentile). The ends of the box show the first (lower) quartile, Q1, and the third (upper) quartile, Q3. Thus a boxplot can show the overall range (maximum – minimum) and the **interquartile range** (IQR = Q3 – Q1). A modified boxplot uses a rule for identifying values that are extraordinary compared to the others (**outliers** or **outside values**). Circles (o) are often used to denote outliers and asterisks (*) to denote extreme outliers if present. Any point below Q1 – (1.5 x IQR) or above Q3 + (1.5 x IQR) is considered an outlier. Extreme outliers are points below Q1 – (2 x IQR) or above Q3 + (2 x IQR). *Box plots can hide gaps and clusters, so they do not show the shape of the distribution.*

Side-by-side Boxplots: These plots are helpful for comparing two or more distributions with respect to the five-number summary. For example, suppose you are interested in comparing the distribution of a variable, such as the salary of the employees of a certain company. If you have information on sex for the group, you might be interested in comparing the distribution of salary of females with respect to males. In this case, the side-by-side boxplot will be an important part of the descriptive analysis of the data set involved.

Bar Charts: One way to display the number or frequency distribution for a categorical variable is with a bar chart. A bar chart shows the percentage of items that fall into each **category** or value of a **categorical variable**. It displays a bar for each category with the height of each bar equal to the number, the proportion, or the percentage of items in that category. If the categories have no inherent order, we could rearrange the bars in the graph in any way we like. In such cases, the shape of the bar graph would have no bearing on its interpretation.

Going back to Application #1, Jane decides that her variable of interest is a categorical variable. Each respondent in the survey belongs to a single category, dependent on the answer they selected. The proper graph to illustrate this categorical variable is a bar chart. The bar chart shown below illustrates that 62% of respondents believe the higher education system was doing either a fair or poor job in their preparation of college graduates.

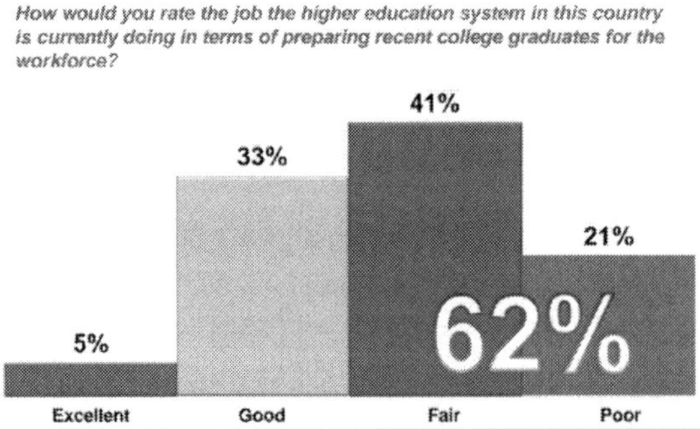

Northeastern_University_Innovation_Imperative_Higher_Ed_Outcomes_Poll_Deck_FINAL_Delivered.pdf

Sequence (Time) Plots: Data might be gathered over time. Employment rate, stock prices, and sales figures are just a few examples. When data is gathered over time, such as the number of runs scored over a one month by the Tigers, it is generally wise to examine the data plotted against time. Plots against time can reveal the main features of a time series, overall patterns and striking deviation from those patterns. Some overall patterns that may arise are:

- A persistent, long-term rise or fall called a **trend** (either increasing or decreasing).
- A pattern that repeats itself at regular intervals of time called **seasonal variation**.
- A persistent, long-term increase or decrease in the **variation** of the observations called a **pattern in variation**.

If data is collected over time, a sequence plot can be used to check the assumption of a random sample, often needed for inference procedures. A random sample consists of *independent* and *identically distributed* (i.i.d.) observations. This means the observations can be considered as all coming from the same parent population (with the same or *identical* distribution) and are *independent* of one other.

With a sequence plot, you can check the *identically distributed* aspect of a random sample by looking for evidence of **stability** in the plot. Stability is supported when both the mean of the observations and the amount of variation among observations appear to be constant over time and there does not appear to be any *pattern* in the resulting plot.

Below are two sequence plots; in the first plot the observations appear to support that the underlying process that generated the observations is stable, but that is not the case for the observations in the second plot on the right. In this case, there appears to be an increasing trend, thus the underlying process does not appear to be stable; the observations should not be considered a random sample.

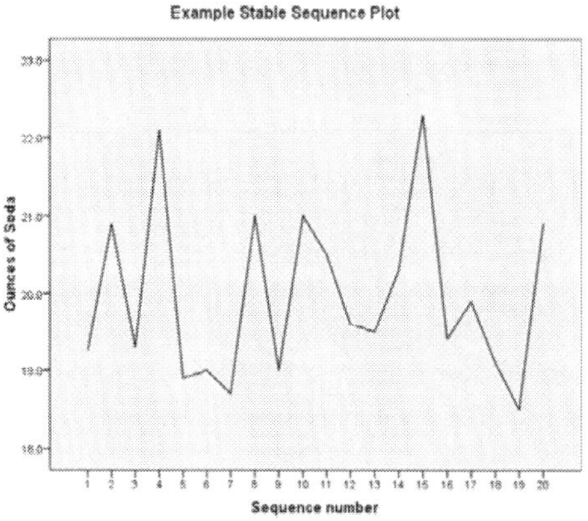

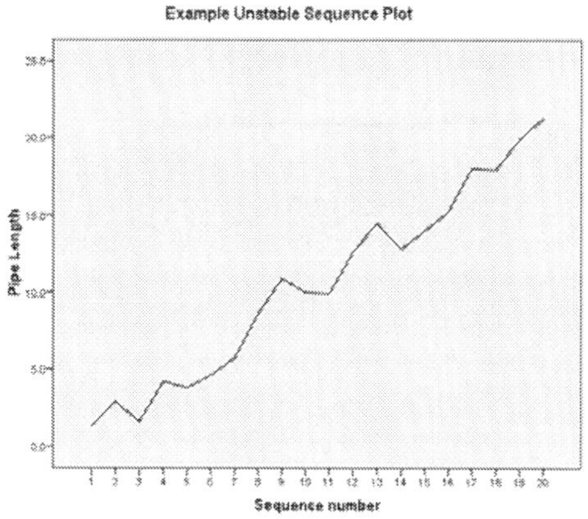

Going back to Application #2, here is the time plot Brad created of the number of runs the Detroit Tigers scored between September 1, 2013 and October 19, 2013. This is another example of an unstable time plot. Brad notes that there is a large variation in the number of runs scored by the Tigers early in September but less variation in their run total around game 21. The underlying process for the number of runs scored does not appear to stable, so Brad should not consider this a random sample.

Example Unstable Sequence Plot

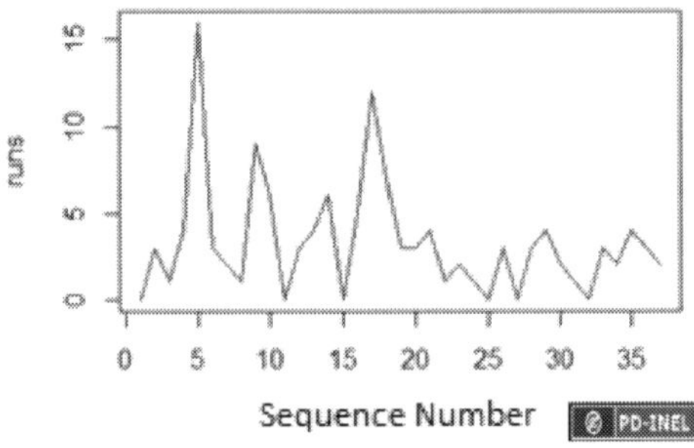

Q-Q Plots: Sometimes the **assumption of a normal model for a population of responses will be needed in order to perform certain inference procedures**. A histogram can be used to get an idea of the shape of a distribution. However, there are more sensitive tools for checking whether the shape is *close* to a normal (bell-curve) model. The *best* plot that can be used to check for normality is called a Q-Q Plot, which plots the percentiles (quantiles) of a standard normal distribution against those of the observed data. If the observations follow an approximately normal distribution, the resulting points should follow roughly a straight line with a positive slope. Strong deviations would indicate possible departures from a normal distribution. At the right is an example of a Q-Q Plot showing data that does seem to come from a population with an approximately normal distribution.

Normal Q-Q Plot of ACT

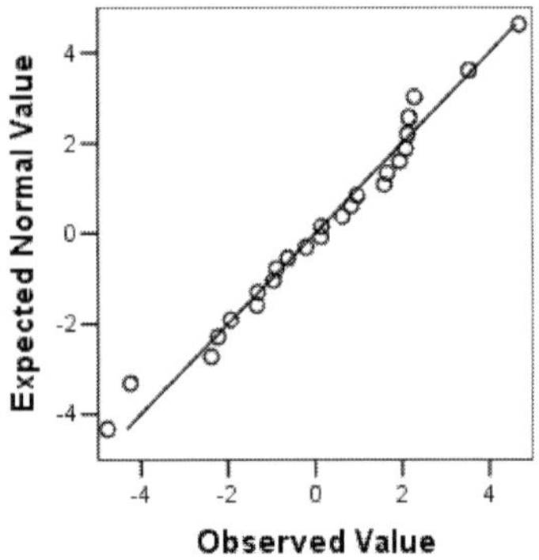

The three graphs below are examples for which a normal model for the response does not seem reasonable. The Q-Q plot on the far left indicates the existence of two clusters of observations. The Q-Q plot in the center shows an example where the shape of the distribution appears to be skewed right. The Q-Q plot on the far right shows evidence of an underlying distribution that has shorter tails compared to those of a normal distribution.

Normal Q-Q Plot of CLUSTER

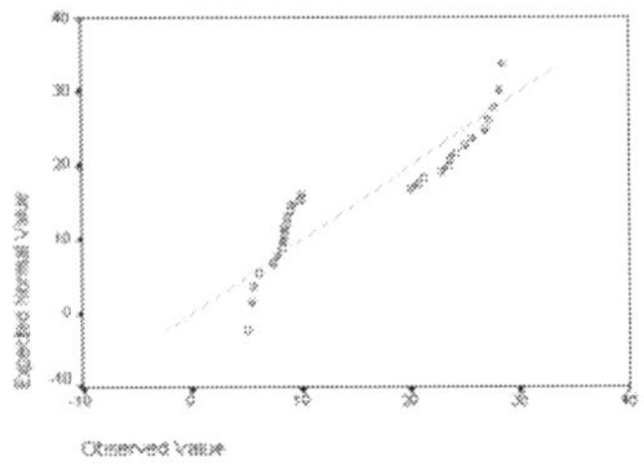

Normal Q-Q Plot of SKEW

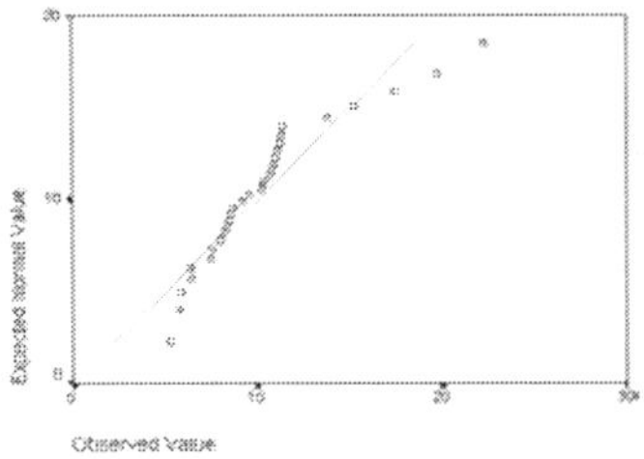

Normal Q-Q Plot of TAILS

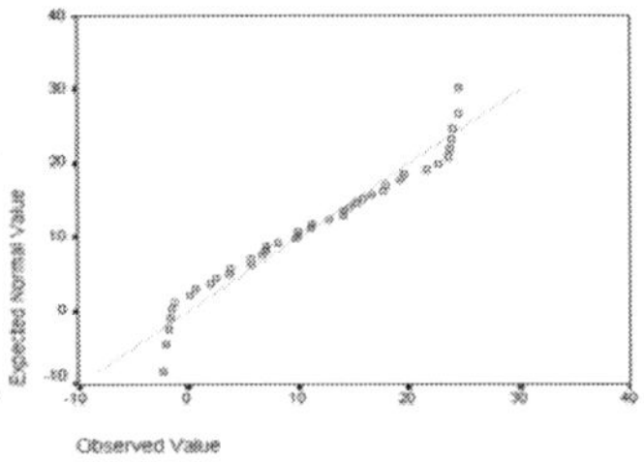

Note: Many inference procedures, including some you will use later in the semester, require the assumption of normally distributed population(s). Most of these procedures are robust which means we would need to see strong evidence of a departure from normality to conclude this assumption is not met. Some mild departures from normality would still allow us to conclude the underlying model for the response is reasonably normally distributed.

Finally, consider the Q-Q plot at the right. In this case, we would say the Q-Q plot shows evidence of an underlying distribution which is approximately normal except for one large outlier that should be further investigated. Outliers could appear in either the upper or lower tail.

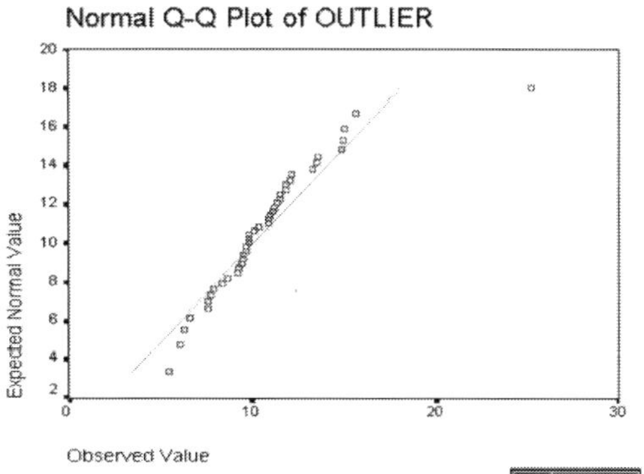

Warm-Up: Matching

Match the graph or descriptive statistic to one of its primary uses.

____ 1. Histogram	A. Measure of center, not sensitive to outliers
____ 2. Bar Chart	B. Compare distributions (but not their shapes)
____ 3. Mean	C. Examine distribution of a categorical variable
____ 4. Median	D. Helps assess if can treat data as a random sample
____ 5. Side-by-side Boxplots	E. Measure of spread
____ 6. IQR	F. Examine distribution of a quantitative variable
____ 7. Time Plot	G. Helps assess if underlying distribution is bell-shaped.
____ 8. Q-Q Plot	H. Measure of center, sensitive to outliers

ILP: More Visualizing and Exploring Quantitative Data

In this In-Lab Project, you will create additional useful graphs and obtain more descriptive statistics for quantitative data using SPSS.

***Task 1:* Boxplots, QQ Plots, and Summaries by group.** Recall the data set **employee data.sav** contains information on employees at a company.

1. Open the **employee data.sav** data set.

2. Obtain a boxplot for current salary. Use: **Graphs> Legacy Dialogs> Boxplot> Simple> Summaries of separate variables**. (This produces a boxplot for one variable with no groups). Click on the button **Define** to open another dialog box that defines the variables for our analysis. Click once on **salary** to highlight it and then on the **Boxes Represent** arrow to select it.

Note: Boxplots do not have a Titles option. However, you may add a title via the Chart Editor. Double click the graph, and from the Chart Editor menus select **Options> Title**. The Chart Editor creates the text box and automatically positions it in the top center of the chart. Type the text and press enter when you are finished typing. To enter line breaks, press Shift+Enter.

   Describe what the boxplot shows about the distribution of current salary. What do the various lines that form the box and the line in the middle of the boxplot represent?

3. Recall from Lab 1 that a histogram of the current salary data indicated a strongly skewed to the right distribution. Let's examine a **QQ plot** to see that it supports this non-normal feature. Create a Q-Q plot for the salary data. Q-Q plots are created via **Analyze> Descriptive Statistics> Q-Q plots**.

   Provide a sketch and explain how the QQ plot supports that the underlying distribution of salary does not appear to be normal (or bell-shaped).

4. You would like to compare the distribution of salary for minorities versus non-minorities. To obtain numerical summaries or any graph (except boxplots) for current salary by minority status, we need to split the data file. Use **Data> Split File** and choose **Organize output by groups**. The grouping variable is minority classification.

> **Note:** Doing this successfully does not produce any noticeable changes in the SPSS windows; there will only be a few short lines of output in the Output window confirming the split.

Create histograms separately for minorities and non-minorities. Compare and contrast the distribution of salary for the two groups. Can we use the same description for both histograms?

5. Obtain descriptive statistics for current salary by minority status. Remember the data is still split, so just generate the descriptive statistics by selecting **Analyze> Descriptive Statistics> Frequencies** and then choose the summary measures you want under the **Statistics** button.

Fill in the basic summary measures for current salary by minority status.

	Minority	**Non-Minority**
Mean:		
Standard Deviation:		
Median:		
IQR = Q3 – Q1:		

6. To create side-by-side boxplots for current salary by minority status, the data file should ***NOT*** be split. Go back to the **Split File** dialog box and choose **Analyze all cases, do not create groups**. Again, the only change will be one line of output confirming the data has been "un-split." Next use **Graphs> Legacy Dialogs> Boxplot** with **Simple** and **Summaries for groups of cases**. Minority Status is the variable for the category axis, and current salary is the variable.

   How does the distribution for current salary compare for minorities versus non-minorities (based on the histograms, descriptives, and side-by-side boxplots)?

Task 2: **Time Plots.** The data set **oldfaithful.sav** contains the date and duration of eruptions (in minutes) of the Old Faithful geyser. The data was collected several times per day over 23 consecutive days.

1. Open the **oldfaithful.sav** data set.
2. Make a time plot for the data using **Analyze> Forecasting> Sequence Charts**. What does the graph show? Are there any patterns to the process?

3. Does the plot appear to be stable?

   Would it be reasonable to conclude these data are a random sample of eruptions?

   Would it make sense to go on to make a histogram of these data? a QQ plot of these data? descriptive summaries? Why or why not?

Cool-Down: Check Your Understanding About ...

... Standard Deviation:

Suppose we are interested in learning about the height of Michigan students. We take a simple random sample of 100 students and find the average height for this sample is 66 inches, with a standard deviation of 2 inches.

A student provided the following statement as an interpretation of standard deviation.

"68% of the height values are within 2 inches of the mean height"

This is not an interpretation of the standard deviation; but the student is making a statement that makes use of the standard deviation. However, you would need to first make a graph to check if this use is appropriate. What graph(s) would you make and what would you look for in the graph?

... Time Plots:

A new method of measuring phosphorus levels in soil is under consideration. A sample of 11 soil specimens is analyzed using the new method. The time series (sequence plot) for the 11 observations is presented.

Comment on the overall stability of these data based on this plot.

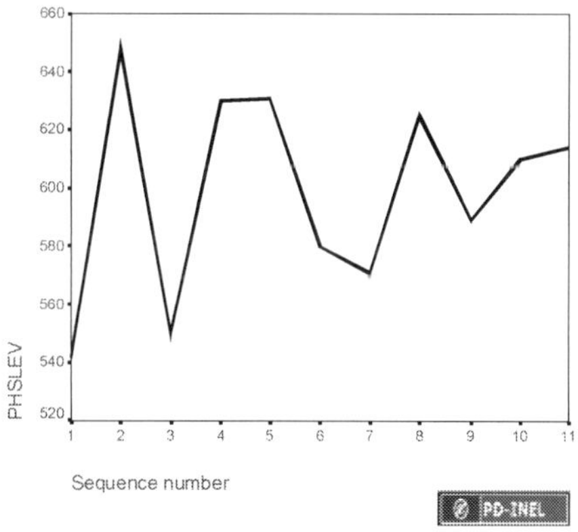

Would it be appropriate to go on to make a histogram of these data? Explain.

Optional Review of Time Plot and QQ Plot Examples

If you would like to look at more examples of QQ plots (which help assess whether the model for the underlying population of responses seems to be normal) and examples of time plots (which help assess whether our underlying process appears to be stable and if we can consider the data to be a random sample) ... try out these tasks using some simulations with R (R scripts).

Task 1: Go to the **Extra Review** link on your course site and click on the **QQ Plots in R** link in the list. Review the background about the simulator and download the **qqplot** script file (which will open up the R program).

1. Begin the program by entering the following command. **qq_plot ()**
2. Select your sample size by entering a number between 1 and 10000.
3. Select if you want a QQ plot from a normal or non-normal distribution.
4. If a non-normal distribution, select the type of distribution you would like to see.
5. Once your QQ plots and the corresponding histograms have been created, you will be asked if you wish to save the plot. You will then be prompted to select a sample size. Create QQ plots for many different samples.
6. Sketch the QQ plot and Histogram for a sample from a skewed right distribution.
   QQ plot: Histogram:

Task 2: Go to the **Extra Review** link on your course site and click on the **Time Plots in R** link in the list. Review the background about the simulator and download the **timeseries** script file (which will open up the R program).

1. Begin the program by entering the following command: **time_series ()**
2. Select your sample size by entering a number between 1 and 10000.
3. Select if you would like an example of a stable or unstable time plot.
4. If unstable time plots are selected, select the type of pattern you want to see.
5. Once your time plots have been created, you will be asked if you want to save the plot. You will then be prompted to select a sample size. Try out the various options and explore the various patterns of time plots.
6. Create a time plot for a sample of 1000 data points taken from a process with an increasing mean and decreasing variance. Sketch the time plot below.

Example Exam Question on Boxplots

Fifty-five parents of grade-school children were recently interviewed regarding the breakfast habits in their family. One question asked was if their children take the time to eat a breakfast (recorded as breakfast status – Yes or No). The grades of the children in some core classes (e.g. reading, writing, math) were also recorded and a standardized grade score (on a 10-point scale) was computed for each child. At the end of the study it was discovered that the children who do take time to eat breakfast get higher grade scores than those who don't.

a. What type of study is this? ***Experiment*** ***Observational study***

b. What is the response variable in this study? ____________________________

c. What is the explanatory variable in this study? ______________________________

d. What type of variable is the explanatory variable? ***Categorical*** ***Quantitative***

Side-by-side boxplots of the children's standardized grade scores are provided.

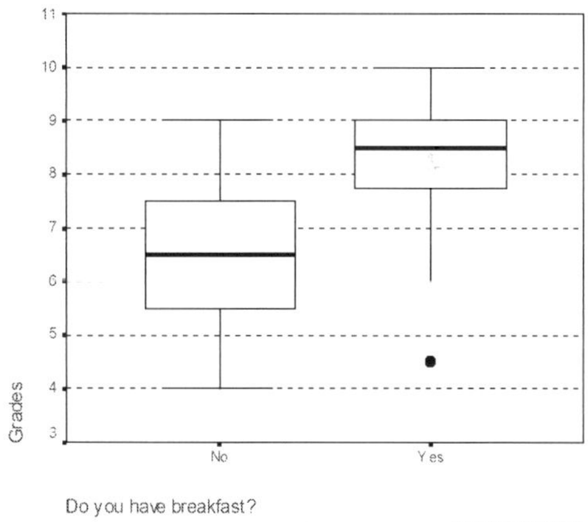

e. What is (approx.) the lowest grade scored by a child who **does** have breakfast?
____________ points

f. What is the IQR for the grade scores of children who **do** eat breakfast?
____________ points

PD-INEL

g. Using a measure displayed in the boxplot, complete the following sentence.

The highest grade scored by one of the children **not** eating breakfast is (approximately) equal to the ________________________ for the children who **do** eat breakfast.

h. True or false: The symmetry in the boxplot for the children **not** eating breakfast *implies* that the histogram for these data will be approximately bell-shaped.
Circle one: ***True*** ***False***
Explain:

i. Which graph would be best for assessing if the underlying distribution of scores (grades) for the children **not** eating breakfast *is* approximately bell-shaped?

Lab 3: Probability and Random Variables

Objective: The objective of this lab is to become familiar with using the models for random variables and to find the probabilities associated with the models you have learned. The probabilities we compute from these models (for example, p-values in testing theories) will help us make reasonable decisions. You will work with three random variables and the methods used to calculate probability for each variable. You will also become familiar with several concepts that allow for easier calculation of probabilities.

Application: Researchers at University of Michigan Hospital and Baylor University were interested in determining if lack of sleep for teenagers led to an increased incidence of heart disease. These researchers conducted a study in which the sleep patterns for 37 teenagers were recorded in addition to the presence of heart disease for each teenager. The researchers can use the probability rules to determine if lack of sleep and incidence of heart disease are independent events.

Overview: In this lab you will be introduced to several random variables and their models. These variables can be classified as one of two types: a ***discrete random variable***, which has a finite number of outcomes, and a ***continuous random variable*** which has an infinite number of outcomes. You will practice computing probabilities for each of these variables, using concepts such as a ***standardized score*** and a ***Normal Approximation.*** There are several situations that can make computing probabilities easier, such as ***independent events*** and ***mutually exclusive events*** that will also be explored in this lab.

Independent Events: Two events A, B are said to be independent if knowing that one will occur (or has occurred) does not change the probability that the other occurs. In probability notation this can be expressed as $P(A|B) = P(A)$. Going back to the application, the researchers from the study can check the independence of the two events by computing the probability of heart disease for teenagers who sleep less than the recommended amount as well as the probability of heart disease for all teenagers. If the incidence of heart disease for a teen is independent of the amount of sleep they get, the two probabilities will be equal.

Mutually Exclusive: Two events A, B are mutually exclusive (or disjoint) if they do not contain any of the same outcomes. So their intersection is empty.

Random Variables: A random variable assigns a number to each outcome of a random circumstance, or, equivalently, a random variable assigns a number to each unit in a population. The distribution of a random variable is a model that shows us what values are possible for that particular random variable and how often those values are expected to occur (i.e. their probabilities). The model can be expressed as a function or table or picture, depending on the type of variable it is. We will consider two broad classes of random variables: discrete random variables and continuous random variables.

Discrete Random Variable: A discrete random variable, X, is a random variable with a finite or countable number of possible outcomes. The probability distribution function (pdf) for a discrete random variable X is a table or rule that assigns probabilities to the possible values of the X.

Two conditions that must always apply to the probabilities for a discrete random variable are:

Condition 1: The sum of all of the individual probabilities must equal 1.

Condition 2: The individual probabilities must be between 0 and 1.

Binomial Random Variable: One type of a discrete random variable is the binomial random variable, which counts the number of times a certain event occurs out of a particular number of observations or trials of a random experiment.

A binomial experiment is defined by the following conditions:

1. There are n "trials" where n is determined in advance and is not a random value.
2. There are two possible outcomes on each trial, called "success" (S) and "failure" (F).
3. The outcomes are independent from one trial to the next.
4. The probability of a "success" remains the same from one trial to the next, and this probability is denoted by p. The probability of a "failure" is $1 - p$ for every trial.

Continuous Random Variable: A continuous random variable, X , takes on all possible values in an interval (or a collection of intervals). The way that we determine probabilities for continuous random variables differs in one important respect from how we determine probabilities for discrete random variables. For a discrete random variable, we can find the probability that the variable X exactly equals a specified value. We can't do this for a continuous random variable. Instead, we are only able to find the probability that X could take on values in an interval. We do this by determining the corresponding area under a curve called the probability density function of the random variable.

So the probability distribution of a continuous random variable is described by a density curve. The probability of an event is the area under the curve for the values of X that make up the event. The probability model for a continuous random variable assigns probabilities to intervals.

Definition: A curve (or function) is called a **Probability Density Curve** if:

1. It lies on or above the horizontal axis.
2. Total area under the curve is equal to 1.

Normal Random Variable: The family of normal distributions is very important because many variables have this shape and form approximately and many statistics that we use in our inference methods are based on sums or averages which generally have (approximately) a normal distribution.

A normal curve is symmetric, bell-shaped, centered at the mean and its spread is determined by the standard deviation. In fact, the points of inflection on each side of the mean mark the values which are one standard deviation away from the mean.

Standardized Scores: A normal distribution is indexed by its population mean, and its population standard deviation. Recall that the standard deviation is a useful "yardstick" for measuring how far an individual value falls from the mean. The standardized score or z-score is the distance between the observed value and the mean, measured in terms of number of standard deviations. Values that are above the mean have positive z-scores, and values that are below the mean have negative z-scores.

Normal Approximation to the Binomial Distribution: The easier way involves using a normal distribution. The normal distribution can be used to approximate probabilities for other types of random variables, one being binomial random variables when the sample size n is large.

Expected Value: The expected value of a random variable is the mean value of the variable X in the sample space, or population, of possible outcomes. Expected value, denoted by E(X), can also be interpreted as the mean value that would be obtained from an infinite number of observations on the random variable.

Standard Deviation: The standard deviation can be viewed as approximately the average distance of the possible values of X from its mean.

Formula Card

Probability Rules

- **Complement rule**

  $P(A^C) = 1 - P(A)$

- **Addition rule**

  General: $P(A \text{ or } B) = P(A) + P(B) - P(A \text{ and } B)$

  For independent events:

  $P(A \text{ or } B) = P(A) + P(B) - P(A)P(B)$

  For mutually exclusive events: $P(A \text{ or } B) = P(A) + P(B)$

- **Multiplication rule**

  General: $P(A \text{ and } B) = P(A)P(B \mid A)$

  For independent events: $P(A \text{ and } B) = P(A)P(B)$

  For mutually exclusive events: $P(A \text{ and } B) = 0$

- **Conditional Probability**

  General: $P(A \mid B) = \dfrac{P(A \text{ and } B)}{P(B)}$

  For independent events: $P(A \mid B) = P(A)$

  For mutually exclusive events: $P(A \mid B) = 0$

Discrete Random Variables

Mean

$$E(X) = \mu = \sum x_i p_i = x_1 p_1 + x_2 p_2 + \cdots + x_k p_k$$

Standard Deviation

$$s.d.(X) = \sigma = \sqrt{\sum (x_i - \mu)^2 p_i} = \sqrt{\sum \left(x_i^2 p_i\right) - \mu^2}$$

Binomial Random Variables

$$P(X = k) = \binom{n}{k} p^k (1-p)^{n-k}$$

where $\binom{n}{k} = \dfrac{n!}{k!(n-k)!}$

Mean

$$E(X) = \mu_X = np$$

Standard Deviation

$$s.d.(X) = \sigma_X = \sqrt{np(1-p)}$$

Normal Random Variables

- $z-\text{score} = \dfrac{\text{observation} - \text{mean}}{\text{standard deviation}} = \dfrac{x - \mu}{\sigma}$
- Percentile: $x = z\sigma + \mu$
- If X has the $N(\mu, \sigma)$ distribution, then the variable

  $Z = \dfrac{X - \mu}{\sigma}$ has the $N(0,1)$ distribution.

Normal Approximation to the Binomial Distribution

If X has the $B(n, p)$ distribution and the sample size n is large enough (namely $np \geq 10$ and $n(1-p) \geq 10$), then X is approximately $N\left(np, \sqrt{np(1-p)}\right)$.

Warm-Up: Types of Variables and Types of Distribution

1. **Type of Variables:**
Today's typical undergraduate student is often characterized as preferring teamwork, experiential activities, and the use of technology. An ECAR (Educause Center for Applied Research) study was published on technology use among undergraduate students. The study used survey and interviewer data to create a portrait of today's students' experiences with and skill using information technology.

Listed below are some of the response variables that were measured in this study. For each of these determine whether it is categorical, quantitative discrete, or quantitative continuous.

a. Technology ownership: Do you own a computer?

   categorical   quantitative discrete   quantitative continuous

b. Time (per week in minutes) spent using a computer for writing documents (word processing).

   categorical   quantitative discrete   quantitative continuous

c. Which social networking site(s) are you a member? (Facebook, Myspace, Friendster, etc.)

   categorical   quantitative discrete   quantitative continuous

---

2. **Types of Models:**
Identify appropriate model for each of the three variables. (Be complete).

a. Below is the model for the random variable X which represents waiting time to be served at a deli.

   X has a ____________________ distribution

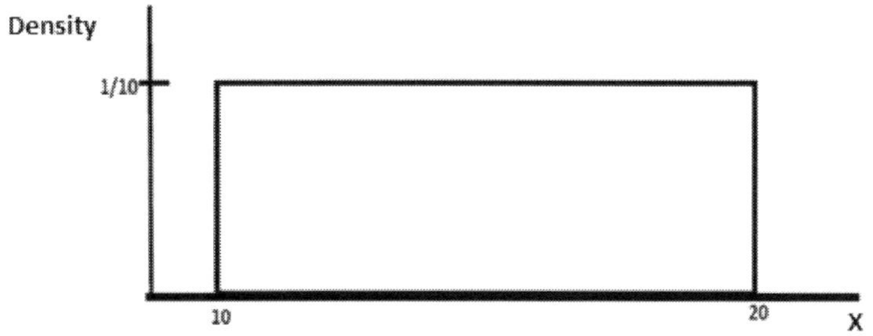

b. Suppose that 45% of Michigan residents own dogs. Let X represent the number of Michigan residents with a dog in a random sample of 10 Michigan residents. A success is defined as the resident owning a dog.

X has a ______________________ distribution

c. Below is the model for the random variable X which represents the score on a Stats 250 Exam (out of 100 points).

X has a ______________________ distribution

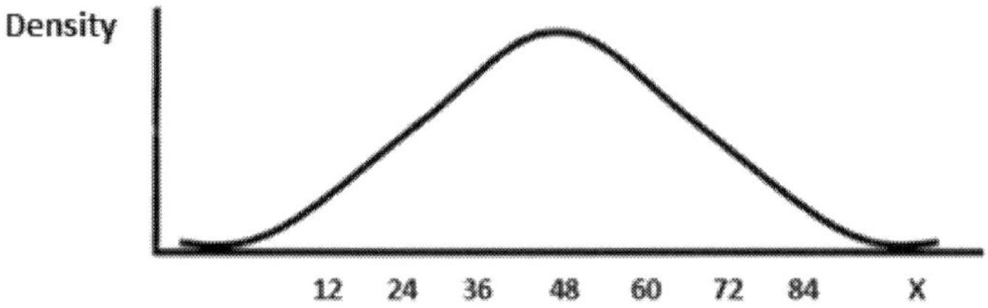

ILP: Probability and Random Variable Review

Problem 1: Study on Smiling

In a recent study people were observed for about 10 seconds in public places (e.g. malls and restaurants) to determine whether they smiled during the randomly chosen 10-second interval. The table shows the results for comparing males (group 1) and females (group 2).

	Smile	No Smile	Total
1 = Male	3269	3806	7075
2 = Female	4471	4278	8749
Total	7740	8084	15824

a. What is the probability that a randomly selected person smiled?

b. The researcher would like to ***assess if smiling status is independent of gender***.

i. To check for independence, the probability found in part (a) should be compared to which of the following probabilities?

P(smiled and male) ***P(smiled given male)*** ***P(male given smiled)*** ***P(male)***

ii. Find the probability selected above and circle the appropriate conclusion.

The probability = ______________________

Thus it appears that smiling status ***is*** ***is not*** independent of gender.

Problem 2: Summer Trip Length

Did high gas prices keep Americans from hitting the road this past summer? In a nationwide survey of adults, one variable measured was how many days vacationers spent driving on the road on their longest trip. Consider the following (partial) probability distribution for the random variable X = the number of days for the longest car trip.

X	4	5	6	7	8
Probability	0.10	0.20	0.25		

a. Suppose the probability of 7 days is twice as likely as the probability of 8 days. Complete the probability distribution for *X*. Show your work.

b. What is the expected number of days for the longest trip?
Include ***symbol, value, and units***.

Problem 3: How Much Time do You Spend Studying Statistics? A Washington Post article "Is college too easy? As study time falls, debate rises" (May 21, 2012) stated that the amount of time college students actually study has dwindled from an average of 24 hours per week to about 15 hours (based on a survey). A professor of statistics decided to ask all of his current semester students to report the number of hours per week they spend studying his course material (on a regular, non-exam week). The mean for the female students was 10 hours and the standard deviation was 3.5 hours.

a. Consider the following interpretations of the standard deviation and clearly circle those that are correct. If the interpretation is incorrect, adjust the statement so it is correct.
   - On average, the number of hours spent studying statistics varied from the mean by about 3.5 hours.
   - The average distance between the number of hours spent studying statistics is roughly 3.5 hours.
   - The average number of hours spent studying statistics is about 3.5 hours away from the mean.

b. Julie is one of his current students and she studies statistics for about 6 hours per week. What is her corresponding z-score?

c. The male students had a lower mean and a larger standard deviation. Suppose Jake's response corresponds to a z-score of 2.1. Is it possible to compare Jake's score with Julie's score?

Problem 4: How Much Time do You Spend Studying Chemistry?

The distribution of hours spent studying (per week) for students in the Chemistry class were reported as being approximately normal with a mean of 12 and a standard deviation of 3.

a. What is the probability a randomly selected Chemistry student studies between 16 and 20 hours per week? First find the z scores, and use the Z-table to estimate the probability. Next, use the prob() R script to get a more exact probability and to produce a nice picture and solution.

b. Jing learns that she is in the top 30% of this distribution. Based on the distribution, Jing must study at least how many hours per week? Make a hand sketch of what you are trying to find to help show your work.

Cool-Down: True or False

Decide whether the following questions are true or false.

1. If the time to wait for pharmacy help has a uniform distribution from 0 minutes to 30 minutes, then 33% of the customers are expected to wait more than 20 minutes. **True** **False**

2. If X has a Binomial (50, 0.7) distribution, then the criteria to use the normal approximation are met. **True** **False**

3. 68% of all test scores will fall within one standard deviation of the mean test score. **True** **False**

4. Police report that 78% of drivers stopped on suspicion of drunk driving are given a breath test, 36% are given a blood test and 22% are given both tests. Do the police administer these two tests independently?
Justify your answer with a calculation. **True** **False**

Example Exam Question on Probability and Random Variables

Suppose that the amount of time spent waiting for your bus to our campus each day is a uniform random variable between 0 to 20 minutes.

a. Sketch a picture of the model for waiting time for the bus. Provide labels for each axis and some values along each axis.

b. What is the probability that you wait at least 10 minutes?

Final answer: ________________

c. What is the probability that you wait at least 10 minutes and that you wait at most 15 minutes?

Final answer: ________________

d. What is the probability that you wait at most 15 minutes, given that you have waited at least 10 minutes already?

Final answer: ________________

Lab 4: Confidence Intervals for a Population Proportion

Objective: This lab will help you better understand the ideas involved in confidence interval estimation as well as how to interpret both the confidence level and confidence interval for a population proportion. You will construct one-sample confidence intervals for a population proportion, and to check that the conditions necessary for the interval are valid.

Application: The University of Michigan wants to learn more about the number of computers that are needed by students on campus as they have the resources to add more computers if needed. We have access to the total number of students that are enrolled at the University and we are asked to conduct a study to assess if more computers are needed. Using the data collected from the study, we can estimate the proportion of all Michigan students who don't have a laptop and will require access to an on-campus computer. The University can use this estimate of the population proportion along with the number of total students on campus to estimate the number of computers on campus that are needed.

Overview: Since this population proportion is an unknown number, we are interested in knowing how close a sample proportion (based on the random sample) is expected to be to the true proportion. When we report the estimate, we should provide both the estimate of UM students who don't own a laptop (sample proportion) and a statement that describes the precision of the estimation process.

Confidence intervals (CIs) provide a method of stating both how close the sample proportion is likely to be to the value of a population proportion and the accuracy of it being that close. The basic structure for any confidence interval is: **estimate ± (multiplier x standard error)**. The "multiplier x standard error" portion is also called the **margin of error** (or error margin). The multiplier used will depend on the confidence level we will use and will be a z multiplier since we are constructing an interval for a population proportion. In contrast, the **confidence <u>level</u>** is the **proportion of times this <u>method</u> *will* produce an interval that contains the true proportion of UM students without a laptop in repeated random sampling** (if this study were to be repeated over and over).

Population Proportion	
Parameter	p
Statistic	$\hat{p}$
Standard Error $\text{s.e.}(\hat{p}) = \sqrt{\frac{\hat{p}(1-\hat{p})}{n}}$	
Confidence Interval $\hat{p} \pm z^{*}\text{s.e.}(\hat{p})$ **Conservative Confidence Interval** $\hat{p} \pm \frac{z^{*}}{2\sqrt{n}}$	
Large-Sample z-Test $z = \frac{\hat{p}-p_0}{\sqrt{\frac{p_0(1-p_0)}{n}}}$	
Sample Size $n = \left(\frac{z^{*}}{2m}\right)^2$	

Confidence Interval (CI) Summary

For Exploring the Population Proportion of UM Students without a Laptop

1. **Interpreting a 95% Confidence Interval Estimate itself:** We are 95% confident that the true population proportion of UM students who do not own a laptop lies inside the CI we will compute. The interval provides a range of reasonable values for this population proportion.

2. **Interpreting a 95% Confidence Level used to make the interval:** If the procedure were repeated many times (that is, if we repeatedly took a random sample of the same size, and computed a 95% CI based on each sample), we would expect 95% of the resulting CIs to contain the true population proportion of UM students who do not own a laptop.

3. **Be Careful about how you use *probability* or *chance*:** The **probability** that the true population proportion of UM students without a laptop is located in a particular, already computed confidence interval is either 0 or 1. Both the interval and the population proportion are fixed entities, so either the true proportion is in that particular interval, or it is not.

4. **How Confidence Intervals can Guide Decision Making:**

   Principle 1: A value in the CI we calculate is an "acceptable" or "reasonable" possibility for the population proportion of UM students without a laptop. A value not in a CI can be rejected as a likely value of this population proportion.

   Principle 2: When the CIs for the proportion for two different populations do not overlap, it is reasonable to conclude that the proportions for the two populations are different.

The above summary and interpretations can be used for any problem by adjusting the confidence level and parameter under study, given the context of the problem. Phrases such as "true population proportion of UM students who do not own a laptop" should be adjusted accordingly.

Warm-Up: Confidence Level Interpretation

Background: A study was conducted to learn about eating habits for American families. A random sample of 200 families was selected, and an adult head of household was asked to complete a survey. One question asked was, "Did your family eat dinner together last Sunday night – yes, no?"

Based on the results, a 95% (conservative) confidence interval for the proportion of all such families that ate dinner together last Sunday night is given by (0.56, 0.70).

Task: As part of an exam question, Stats 250 students were asked to write a concise statement to explain the meaning of the confidence level of 95% (*'interpret the confidence level)*. Below are some of the answers received. For each, determine if it is a correct or an incorrect interpretation of the meaning of the confidence **level**.

1. If this survey was repeated many times, we are confident the true population proportion of families that ate dinner together last Sunday night is represented in the data 95% of the time.

   Correct **Not Correct**

2. The 95% confidence level means that with this method, and for similar samples, we would construct many confidence intervals, and of these, 95% would contain the true population proportion of families that ate dinner together last Sunday night.

   Correct **Not Correct**

3. In the long run, the population proportion of families that ate dinner together last Sunday night will be found in the interval 0.56 to 0.70, 95% of the time.

   Correct **Not Correct**

4. In data collected from a normal distribution, 95% of the observed values will lie within two standard deviations of the mean.

   Correct **Not Correct**

ILP: Calculating a CI for a Population Proportion

In this ILP, we are going to investigate using the sample proportion as a way of estimating the population proportion. For the purpose of this activity, we are going to assume that the lab section is a representative random sample of the UM student body. Follow the steps below to estimate the population proportion.

1. Determine a question to investigate by filling in the following question.

   "What proportion of UM student ...

   ____?"

2. Once the question has been determined, the instructor will pose the question to the class using clickers. Record the results below and provide an estimate of the population proportion of all UM students that ________________ (that is, report the sample proportion).

# of students responding:	
# of student who responded "yes":	
Sample proportion of student who responded "yes" : (symbol and value)	

3. (a) The estimate in part (2) is a sample proportion. About how far away from the population proportion would you expect such estimates to be, on average?

   (b) There are certain conditions required for this inference procedure to be valid. Are they met here? If not, how would you update this study so the conditions would be met?

(c) Use your results to construct a 95% confidence interval estimate for the population proportion of all UM students that ______________.

(d) What would happen to the width of your interval if you were to construct a 99% CI instead (using the same survey results)?

(e) Interpret the **interval** you constructed in part (b).

(f) Explain what the corresponding confidence **level** means.

(g) What sample size would be needed for a 95% (conservative) confidence interval with a margin of error of 0.06 (or 6%)?

Cool-Down: Interpretations of Confidence Intervals

Consider once again the survey of American families, and recall that the 95% (conservative) confidence interval for the proportion of all such families that ate dinner together is given by (0.56, 0.70).

Below are answers provided by students when asked to now give an interpretation of the actual 95% confidence interval computed in the study. Assess which are correct and which are not correct explanations as to what the interval (0.56, 0.70) means (in the context of this study)?

1. There is a 95% chance the confidence interval of (0.56, 0.70) will contain the true population proportion of all families that ate dinner together last Sunday night.

   Correct **Not Correct**

2. We estimate with 95% confidence that the true population proportion of all families that ate dinner together last Sunday night will be found in the interval 0.56 to 0.70.

   Correct **Not Correct**

3. If we repeated this study many times, the true proportion of all families that ate dinner together last Sunday night will be in the interval (0.56, 0.70) about 95% of the time.

   Correct **Not Correct**

Example Exam Questions
on Confidence Intervals for a Population Proportion

1. **Point Estimate and Standard Error**

   After once again losing a basketball game to their archrival, a college's alumni association conducted a survey to see if alumni were in favor of firing the coach. In a random sample of 100 alumni, 70 were in favor of firing the coach.

   a. Report the sample proportion of alumni in favor of firing the coach and the corresponding standard error.

   b. Which is/are correct interpretation(s) of the standard error in part (a)?

      i. If repeated samples of 100 alumni were obtained, we would estimate the resulting sample proportions to be about 0.046 from the true population proportion p on average.

      ii. If repeated samples of 100 alumni were obtained, we would estimate the sample proportion of 0.70 to differ from the true population proportion p by about 0.046.

      iii. We would estimate that our sample proportion of 0.70 is about 0.046 away from the true population proportion p on average.

2. **Construct a CI for population proportion *p* and Find the Sample Size.**

a. Provide a (general) 95% confidence interval estimate for the population proportion of all alumni who are in favor of firing the coach.

b. Write a sentence to interpret the 95% confidence **interval** that you computed in part (a).

c. Interpret the meaning of the 95% confidence **level** used in making the interval.

d. Give a **conservative** 95% confidence interval estimate for the population proportion of all alumni who are in favor of firing the coach.

e. Is the conservative interval in part (d) wider or narrower than the interval in part (a)?

f. What sample size would be needed for creating a 95% (conservative) confidence interval with an error margin of just 0.05?

3. Point Estimate and Interpretation of the Confidence Level

The 99% confidence interval for the p, the population proportion of all club members who participated in the recent fundraiser, is given by (0.30, 0.40).

a. The above interval was constructed from the results of a random sample of $n = 200$ club members.

   i. What was the proportion of club members in the sample that participated in the fundraiser?

   ii. How many club members in the sample participated in the fundraiser?

b. Does the 99% confidence level imply that $P(0.30 \leq p \leq 0.40) = 0.99$?

   Yes ***No***

   Explain.

Lab 5: Hypothesis Testing for a Population Proportion

Objective: In this lab, you will learn an important statistical technique that will allow you to answer the question, *"Was our observation due to chance, or is it more significant?"* The objective is to guide you through the ideas behind tests of statistical significance and the statistical language involved. This lab first presents a general overview of testing. Then, the In-Lab Project discusses the large sample Z-test for a population proportion, as well as providing practice in answering the question of whether an event can be attributed to chance.

Application: Lucy is an Ann Arbor resident who is considering the possibility of running for a seat on the City Council. To decide if she should invest time and money in her campaign, she wants to conduct some research to see if there is evidence that she could win the election (i.e. she could receive a majority of all the votes).

Overview:

Hypothesis Tests: A **test of hypotheses** (or **significance test**) is a procedure designed to assess what the evidence provided by the data says about some statement about a population parameter. When we conduct this hypothesis test for Lucy, we are interested in testing if the population proportion of votes that would be cast for Lucy is greater than 0.5, ensuring her victory.

Hypothesis Testing Steps:

1. Determine appropriate null and alternative hypotheses and set α.
2. Check assumptions for performing the test.
3. Calculate the test statistic and determine the *p*-value.
4. Evaluate the *p*-value and report a conclusion in the context of the problem.

The first step of this hypothesis test is to identify the hypotheses; this step is crucial, as it dictates the procedures for the remainder of the test. The **null hypothesis**, H_0, represents the status quo or statement of no effect. The **alternative hypothesis**, H_a (or sometimes H_1), represents the experimenter's new model, or what the experimenter would like to show.

For our application regarding Lucy, we could define *p* to represent the true proportion of all Ann Arbor residents who would vote for Lucy. We want to test the hypothesis H_0: $p = 0.5$ against the alternative H_a: $p > 0.5$.

Both of the hypotheses are postulated about the same **population** proportion. The alternative hypothesis can take three different forms – it may be a denial of the null hypothesis (uses$\neq$; called a **two-sided test**), or it may specify a direction of interest (uses > or <; called a **one-sided test**). Notice that we <u>never</u> test for equality in H_a.

The purpose of a significance test is to assess whether or not the observed data are consistent with the null hypothesis (within the reasonable bounds of sampling variability). If we collect our sample and compute a population proportion of votes for Lucy that is much higher than the 0.5 we specified in the null hypothesis, we have evidence to reject that null hypothesis. To help us make this decision, we use a **test statistic**, which represents a summary of the data. When we conduct a test about a population proportion, we are performing a Large Sample Z test. The test statistic for this Large Sample Z test has the following form: $z = \frac{\hat{p}-p_0}{\sqrt{(p_0)(1-p_0)/n}}$.

This p_0 is the value specified in the null hypothesis. For Lucy's test: $p_0 = 0.5$. The sample proportion of people who will vote for Lucy is our $\hat{p}$. The test statistic tells us how many standard errors the sample proportion, $\hat{p}$, is from the test value, p_0.

The test relies on two key assumptions: that we have a random sample (our sample of responses is representative of the large population of all such responses), and that our sample size is large enough. To achieve a large enough sample size, we need at least 10 hypothesized yes answers and at least 10 hypothesized no answers, or said another way, under the null hypothesis, would we expect to see at least 10 yes answers and at least 10 no answers.

Conditions: np_0>10 and $n(1 - p_0)$>10

The Z test statistic has a known probability distribution (under the null hypothesis), and will be examined for evidence in favor of or against H_0. Under the null hypothesis that the true proportion of votes for Lucy is equal to 0.5, the test statistic Z has a standard normal distribution.

In hypothesis testing, another frequently reported value is the **p-value**, a number that is used to indicate the degree of significance of the data. *The **p-value** is the probability of getting a **test statistic** as extreme or more extreme than the observed value of the **test statistic**, assuming the null hypothesis is true.* For Lucy's test, the p-value is the probability of getting a z test statistic like we did or greater, assuming that the true proportion of votes for Lucy is equal to 0.5.

We must decide in advance how much evidence against H_0 we will require for rejection. This designated amount of evidence is called the **level of significance**, denoted by α (alpha). Common values of α are 0.01, 0.05, and 0.10. If the *p*-value is less than or equal to α, we make the decision to reject H_0. For Lucy's test, if we encounter a p-value smaller than the significance level, we would decide to reject the hypothesis that the true proportion of votes for Lucy is exactly 0.5. If we reject the null hypothesis, the results of the test are said to be **statistically significant** at level α. A "significant" result in the statistical sense does not necessarily imply an "important" result in the practical sense. It simply means that such a difference from the null hypothesis is not likely to happen just by chance.

Of course, no procedure is perfect, and as such, there are two types of errors possible during hypothesis testing. If the null hypothesis is true but the decision is to reject H_0, then we say that a **Type I error** has occurred. If we commit a Type I error for Lucy's test, we would reach the decision that Lucy will win the election by receiving more than 50% of the votes, when in fact the true proportion of votes cast for Lucy is 0.5. Lucy would invest her time and money in her campaign when in fact she will not win the election. A **Type II error** occurs when the alternative hypothesis is true, but we fail to reject H_0. A Type II error could be committed in Lucy's test if we decide Lucy will lose the election by receiving 50% of the votes or less, when in fact Lucy will receive greater than 50% of the votes and win the election. Lucy would decide not to run in the election when she would have won a seat on the City Council. Each type of error has a probability of occurring. If the null hypothesis is true, the level of significance, α, is also the probability of a Type I error, while the probability of a Type II error is denoted by β.

Truth	Decision Made	Result	Associated Probability
H_0 True	Reject H_0	Type I Error	α
	Do Not Reject H_0	Correct Decision	$1-\alpha$
H_a True	Reject H_0	Correct Decision	$1-\beta$ = power
	Do Not Reject H_0	Type II Error	β

Another important component of a hypothesis test is its **power**. The **power** of a test measures its ability to detect an alternative hypothesis when it is true. Power of a particular test is calculated as the probability that the test will reject H_0 when the alternative hypothesis is true.

For Lucy's test, the power is calculated as the probability that we conclude Lucy will receive more than ½ the votes in the election when the truth is that she actually will win the election. Since we just learned that β is the probability that we did NOT reject H_0 when the H_a is true, we can see that power is represented by $1-\beta$.

Formula Card:

Corresponding details and formulas as they appear on the Stats 250 formula card are shown at the right.

Population Proportion

Parameter p

Statistic $\hat{p}$

Standard Error

$$s.e.(\hat{p}) = \sqrt{\frac{\hat{p}(1-\hat{p})}{n}}$$

Confidence Interval

$$\hat{p} \pm z^* s.e.(\hat{p})$$

Conservative Confidence Interval

$$\hat{p} \pm \frac{z^*}{2\sqrt{n}}$$

Large-Sample z-Test

$$z = \frac{\hat{p}-p_0}{\sqrt{\frac{p_0(1-p_0)}{n}}}$$

Sample Size

$$n = \left(\frac{z^*}{2m}\right)^2$$

Warm-Up: Stating the Hypotheses and Defining the Parameter of Interest

For each example, fill in the hypotheses and define the parameter of interest.

1. The Detroit Tigers advertising team believes that about 70% of Ann Arbor residents will attend a Tigers game this season. Suppose the general manager wants to cut back on advertising in the area and speculates that Ann Arbor residents are less likely to attend a Detroit Tigers game this season than previously assumed.

   H_o:________________________ H_a:________________________

   Let the parameter ____ represent

2. The University of Michigan is interested in examining the proportion of students who graduate in four years. Suppose that a researcher speculates that the University has a majority of students who graduate four years after they enroll.

   H_o:________________________ H_a:________________________

   Let the parameter ____ represent

ILP: Hypothesis Testing for a Population Proportion

In this ILP, we are going to use a sample proportion to test a theory about the value of the population proportion. Again, we assume the lab section is a representative random sample of the UM student body.

1. Determine a question to investigate by filling in the following question.

   "What proportion of UM students ...

   ____?"

2. Once the question has been determined, the instructor will pose the question to the class using clickers. Record the results below and provide an estimate of the population proportion of all UM students that ________ (that is, report the sample proportion).

# of students responding:	
# of student who responded "yes":	
Sample proportion of student who responded "yes" : (symbol and value)	

3. The estimate in part (2) is a sample proportion. About how far away from the population proportion would you expect such estimates to be, on average? (i.e. report the standard error of the sample proportion)

4. We wish to use the data collected to assess:
"Do a ... (select one) ***minority*** ***majority*** of UM students____________?"
Express this as an appropriate null and alternative hypothesis:

H_0:____________________________________

H_a:____________________________________

where ***p*** represents:____

5. A large sample Z test can be performed using a normal approximation for the binomial if the sample size is large enough. Provide the checks necessary to see if this approximation may be made.

6. Assume the sample is large enough and calculate the corresponding z test statistic value. Show your work.

What is the distribution of the test statistic under the null hypothesis?

7. Use your test statistic to find the corresponding *p*-value. First make a quick sketch by hand and use Table A.1 to find your *p*-value; then try the new pval() script in R to check your answer.

What would the p-value look like if the alternative hypothesis had been that the population proportion did not equal 0.5?

8. What is your decision at the 5% significance level? **Reject H_0** **Fail to Reject H_0**

Also write out your real world conclusion in the context of the problem.

Cool-Down: Errors and Power Computation

Maize and Blue Marbles: Cory is going to play the game 'What is in the Box?'

He will win season tickets to the Red Wings if he can correctly identify the contents of a box (which he cannot see). There are two possibilities for the box contents:

H_0: The box contains eight red marbles and four white marbles.

H_a: The box contains two red marbles and ten white marbles.

Cory will select one marble from the box without looking. He must make his decision based on the color of that one selected marble. He has picked the following decision rule to use: **Reject H_0 if the observed color of the selected marble is WHITE.**

It may help to make a quick visual picture of these two boxes.

a. For this situation, what is the probability of incorrectly rejecting H_0?

2/12   4/12   10/12

b. For this situation, what is the probability of incorrectly failing to reject H_0?

2/12   4/12   10/12

c. For this situation, what is the power?

2/12   4/12   10/12

Example Exam Question: Testing for a Population Proportion

Obama versus McCain: A mock presidential election was planned on the UM college campus before the actual election was held. An enterprising student planned to use the results of the mock election to test the hypothesis that Obama would have a majority of votes among all UM college students. The significance level of the test is set at 5%. A random sample of 3000 college students at the UM was selected to take part in this mock election which resulted in 1578 voting for Obama.

a. State the hypotheses to be tested and completely define the parameter.

H_0:______________________ H_a:______________________

where the parameter ____ is defined (in the context of this problem) as ...

b. Provide the value of the test statistic (including its symbol) and the corresponding *p*-value. Include a sketch to show how the *p*-value is found.
Sketch (include labels):

Symbol ↓

Test Statistic = ________ = ______________

p-value = __________________________

c. Give the decision and conclusion by circling the appropriate statements:

The statistical decision at a 5% significance level is: **Reject H_o** **Fail to reject H_o**

Therefore, there **is** **is not** sufficient evidence to say that, a majority of all students (represented by this sample) would vote for Obama.

d. If this decision in part (c) is incorrect, it would be a: **Type 1** **Type 2** error.

Lab 6: Sampling Distributions and the CLT

Objective: The objective of this lab is to give you a hands-on discussion and understanding of sampling distributions and the Central Limit Theorem (CLT), a theorem that plays an important role in statistics. In this lab, you will simulate random samples from a known population distribution and compute a sample statistic for each of the generated samples. The generated sample statistics can be examined to learn about properties of the sampling distribution of the statistic.

Application: Jack is a researcher who wants to perform a hypothesis test to learn something about the mean salary for all Major League Baseball players. This test requires that the distribution of salaries have a normal distribution. When Jack creates a histogram of the salaries for a sample of players he notices that this distribution is not normal. However, all hope is not lost for doing this hypothesis test. The distribution of the sample mean has a nice feature in which the sample mean will have a normal distribution if some conditions are met.

Overview: Statistical inference is the process of drawing conclusions about a population parameter based on data. When a sample is selected from a population, a summary number can be computed from the observations resulting in the value of a **statistic**. *A statistic is used to estimate the corresponding value for a population (that is, a sample statistic estimates a population parameter).* However, a sample chosen at random will not necessarily yield an estimate (or statistic) that is exactly equal to the corresponding parameter for the population; the next selected sample of the same size will probably give a different estimate from the first one. If additional samples of the same size were taken, you would begin to see how the possible estimates (possible values of the statistic) vary and how close they tend to be to the parameter value.

With a large number of samples, you can assess whether the value of the statistic (e.g., sample mean, $\overline{X}$) will frequently be close to the true value of the population parameter (e.g., population mean, μ), and if so, how close on average. This can be seen more easily through some pictures:

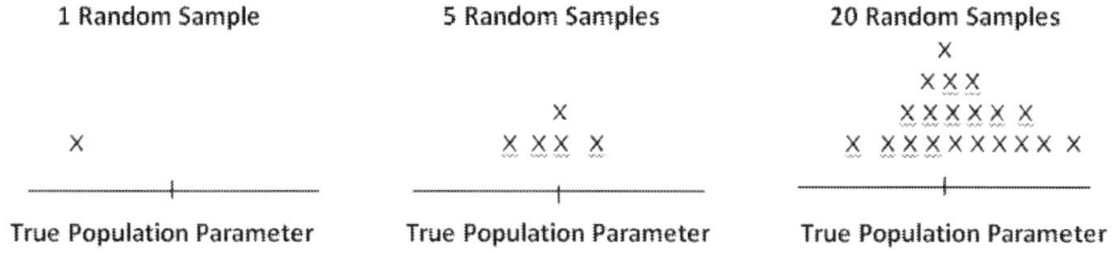

Note: Each X represents one statistic value (one estimate) computed from one sample.

When data are gathered by random sampling, the statistic will be a random variable and as such, it will have a probability distribution. The probability distribution of the sample statistic is called its **sampling distribution**.

Generally, if we use a statistic to make an inference about a population parameter, we want its sampling distribution to be centered at the true parameter (a characteristic which allows us to call that statistic **unbiased**), and we would like variability in the estimates to be as small as possible.

Below, we have two estimators that are both unbiased, but Estimator I has less variability (is more precise). Thus, we would prefer Estimator I to Estimator II.

We will next examine the sampling distribution of the sample statistic most commonly used for measuring the center of a distribution -- the sample mean.

Formula Card:

Sample Means

Mean

$E(\bar{X}) = \mu$

Standard Deviation

$s.d.(\bar{X}) = \dfrac{\sigma}{\sqrt{n}}$

Sampling Distribution of $\bar{X}$

If X has the $N(\mu, \sigma)$ distribution, then $\bar{X}$ is $N\left(\mu, \dfrac{\sigma}{\sqrt{n}}\right)$.

If X follows *any* distribution with mean μ and standard deviation σ and n is large,

then $\bar{X}$ is approximately $N\left(\mu, \dfrac{\sigma}{\sqrt{n}}\right)$.

This last result is **Central Limit Theorem.**

Warm-Up: Check Your Understanding

Suppose the time to wait for placing an order at a drive-through window has a uniform distribution between 0 and 8 minutes. The mean waiting time is 4 minutes and the standard deviation is 2.3 minutes.

Suppose a random sample of 100 drive-through window orders will be selected from over the past week and the average waiting time for placing the orders will be calculated. Which graph below best represents the model that could be used to approximate the probability that the average waiting time is at most 5 minutes?

Graph A

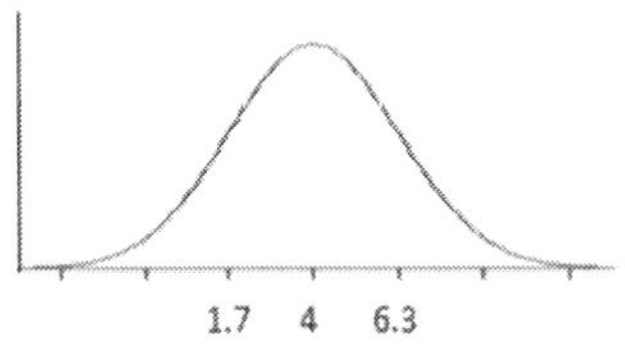

Graph B

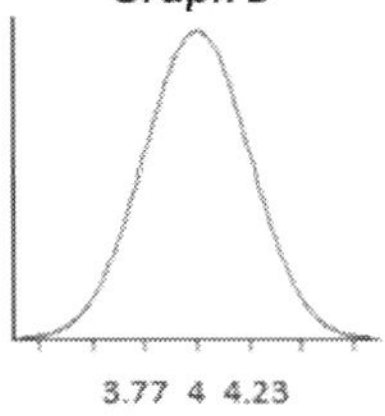

Graph C

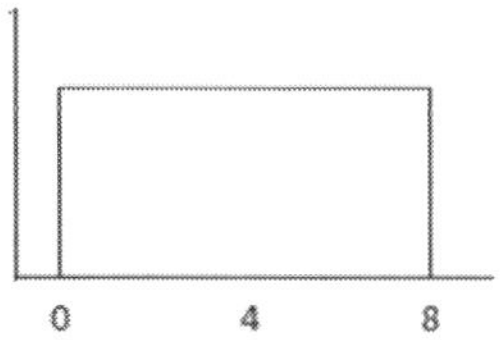

ILP: How Sample Size and Distribution of Parent Population Affect the Sampling Distribution of the Sample Mean

In this activity, you will observe the effects that sample size and the distribution of the population you are sampling from have on the sampling distribution of the sample mean. The sampling distribution of the **sample mean**, $\overline{X}$, is the distribution of the sample mean values for all possible samples of the same size from the same population.

Open the **Sampling Distribution applet** which can be found at:
http://onlinestatbook.com/stat_sim/sampling_dist/index.html

This applet will help you simulate sampling distributions for a variety of statistics and allows you to vary the sample size and the population from which the samples are taken.

Read the Applet instructions.

Press **Begin** to open the applet; you will see the screen pictured below.

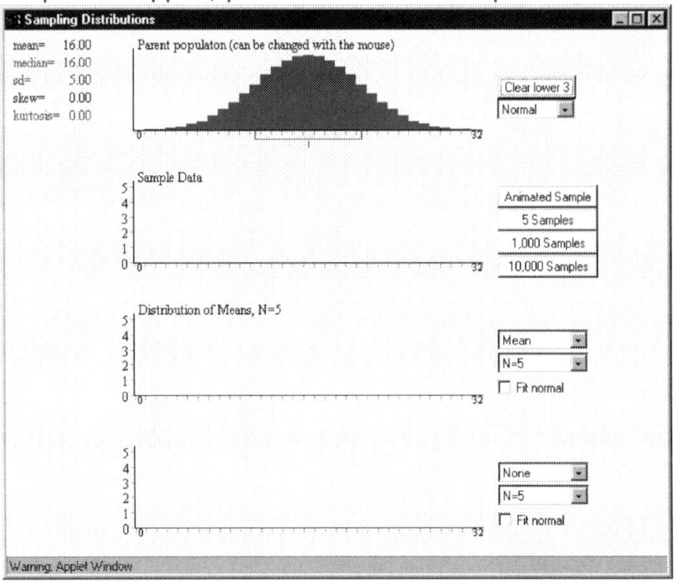

Notice that when the applet begins, a histogram of the normal distribution with mean 16 and standard deviation 5 is displayed for the default parent distribution.

The Sampling Distribution Applet has several options from which you can choose:

- The 1st histogram, the **Parent Population** histogram, is the population from which the sample will be drawn. You can select from **Normal**, **Uniform**, **Skewed**, or even customize the distribution by selecting **Custom** and dragging the mouse over the plot. For now, keep the default N(16, 5) distribution as the parent population. When you are done with a particular simulation, you can click on **Clear lower 3** button to clear the remaining histograms, and select new settings for your next simulation.

- The 2nd plot, the **Sample Data** histogram, displays a histogram of the sampled data. This histogram is initially blank. You can select to draw **Animated Sample**, **5 Samples**, **1,000 Samples**, or **10,000 Samples** from the parent population.

- The 3rd and 4th histograms show the distribution of statistics computed from the sampled data. The number of samples (replications) on which the 3rd and 4th histograms are based is indicated by the label "Reps=," which is displayed once the simulation is started. You can also control which statistic to examine, as well as the sample size by using the drop-down menu options to the right of each plot. (Note that the applet uses N to denote sample size, whereas we generally use *n*.)

The statistic options include:

- **Mean**
- **Median**
- **sd** = standard deviation (uses N in the denominator)
- **Variance** = variance of the sample (uses N in the denominator)
- **Variance (U)** = unbiased estimate of variance (uses N-1 in the denominator)
- **MAD** = mean absolute value of the deviation from the mean
- **Range**

Select **Mean** as the statistic in the 3rd histogram and a sample size of 5 (default), then click on **Animated Sample** to draw one sample of size $n = 5$ from the normal parent population. You will see five observations appear in the 2nd histogram, and the sample mean of the five numbers will appear in the 3rd histogram as a blue rectangle. This graphically shows the process of attaining the sample mean from one sample of size 5. Repeat this several times and you will see how the *sampling distribution of the sample mean* starts to form in the 3rd histogram. Once you have a feeling of this works, you can speed things up by choosing the larger sampling options, such as 5, 1,000, or 10,000 samples.

1. Select the **Normal** distribution as a parent population.
   a. What are the mean and standard deviation of this population?

   b. Select **Mean** (sample mean) as the statistic of interest in both the 3rd and 4th histograms, sample size ***n* = 5** for the 3rd histogram, and ***n* = 25** for the 4th. Do about 5 animated samples, and then take 10,000 samples at once. Draw rough sketches of each of the distributions of the sample means. Make sure to label both axes.
   $n = 5$: $n = 25$:

   c. How do the distributions of each sample mean in the 3rd and 4th histograms compare with the parent population in the 1st histogram? Comment on shape, mean, standard deviation, etc.
   $n = 5$: $n = 25$:

d. Looking at the properties of the population and sample distributions (displayed to the left of their respective histograms), what can you say about the relationship between the standard deviation of the sample mean and the population standard deviation?

e. What can you say about the relationship between the sample size and the standard deviation of the sample mean?

f. Does the number of replications influence the shape of the sampling distribution? That is, as you take more samples, does the shape of the sampling distribution change significantly?

2. Clear the lower three graphs and then select the **Skewed** distribution as a parent population.

a. Select **Mean** (sample mean) as the statistic of interest in both the 3rd and 4th histograms, sample size ***n*** **= 5** for the 3rd histogram, and ***n*** **= 25** for the 4th. Do about 5 animated samples, and then take 10,000 samples at once. Draw rough sketches of each of the distributions of the sample means. Make sure to label both axes.

$n = 5$: $n = 25$:

b. How do the distributions of each sample mean in the 3rd and 4th histograms compare with the parent population in the 1st histogram? Comment on shape, mean, standard deviation, etc.

$n = 5$: $n = 25$:

c. How do the distributions of each sample mean in the 3rd and 4th histograms compare to each other? Comment on shape, mean, standard deviation, etc.

$n = 5$: $n = 25$:

d. How do the distributions of each sample mean in the 3rd and 4th histograms compare with those created of the sample mean when the parent population was normal (in question 1)? Comment on shape, mean, standard deviation, etc.

$n = 5$: $n = 25$:

e. What should be the value of the standard deviation of the sample mean if the population standard deviation is 6.22 and the sample size is $n = 25$? (Show the calculation.) How does this value compare to the standard deviation displayed to the left of the 4th histogram created above?

3. Clear the lower three graphs and select the **Custom** distribution as a parent population. The parent population plot should be empty. To create a distribution, you will need to use the mouse to click and drag on different parts of the parent population graph until you've drawn a distribution that you like. Try and make the distribution as unusual as possible!

a. Provide a rough sketch of your custom population. Be sure to note the mean and standard deviation.

b. Select **Mean** (sample mean) as the statistic of interest in both the 3rd and 4th histograms, sample size ***n* = 5** for the 3rd histogram, and ***n* = 25** for the 4th. Do a few animated samples, and then take 10,000 samples at once. Draw rough sketches of each of the distributions of the sample means. Make sure to label both axes.

$n = 5$: $n = 25$:

c. How do the distributions of each sample mean in the 3rd and 4th histograms compare with the parent population in the 1st histogram? Comment on shape, mean, standard deviation, etc.

$n = 5$: $n = 25$:

d. How do the distributions of each sample mean in the 3rd and 4th histograms compare to each other? Comment on shape, mean, standard deviation, etc.

$n = 5$: $n = 25$:

e. Considering the changes observed from $n = 5$ to $n = 25$ in questions 2 and 3, what can you say about the shape of the distribution of the sample mean with respect to the sample size n?

f. What should be the standard deviation of the sample mean for samples of size $n = 25$ from your custom population? (Show the calculation.) How does this value compare to the standard deviation displayed to the left of the 4th histogram created above?

Cool-Down: Check Your Understanding about the CLT

1. **If the parent population is NOT a normal distribution**, but with a mean μ and a standard deviation σ, then for **a large sample size**, the **sample mean** will have ***approximately*** a ____________________ distribution with a mean of ___________ and a standard deviation of ______________.

This result is known as the **Central Limit Theorem**.

2. Fill out the chart below to further summarize your findings regarding the sampling distribution of the sample mean based on the CLT.

Sample Settings	**Will the sampling distribution of sample mean be approx normal?**
n **= 10, Parent Population Normal**	
n **= 10, Parent Population NOT Normal**	
n **= 50, Parent Population Normal**	
n **= 70, Parent Population NOT Normal**	

Example Exam Question on Sampling Distribution of the Sample Mean

For a particular community it is known that the mean amount of water used per home during October is 1250 gallons and the standard deviation is 325 gallons.

a. The distribution for amount of water used is **skewed to the right**. Sketch a skewed right distribution below and label both axes.

b. For a promotional campaign, a radio station plans to randomly select 50 homes and pay their water bills for the month of October. Describe the **approximate sampling distribution of the sample mean amount of water used for a random sample of 50 homes**. Provide **all** features of the distribution.

c. The radio station can afford to pay for a total of 67,000 gallons. What is the **probability** that the total number of gallons for a random sample of 50 homes will exceed 67,000 gallons?
Hint: Think about how a total and an average are related.

Lab 7: One-Sample *t*-Test Procedures

Objective: In this lab, you will learn an important statistical technique that will allow you to answer the question, ***"Was our observation due to chance, or is it more significant?"*** The objective is to guide you through the ideas behind tests of statistical significance and the statistical language involved. This In-Lab Project discusses the one-sample *t*-test for a population mean, as well as providing practice in answering the question of whether an event can be attributed to chance.

Application: Derek is considering attending the University of Michigan and he wants to consider the cost of rent in Ann Arbor to help make his decision. Ann Arbor's website claims that the average price for apartments in Ann Arbor is $800 a month. However, Derek's friends lead him to believe that the average price for all apartments in Ann Arbor is actually greater than $800 a month. Derek would like to conduct a hypothesis test to see if his friends' statement is supported.

Overview:

Hypothesis Tests: A **one-sample t test** is a procedure designed to assess what the evidence provided by the data says about some statement about the population mean. The key elements of the one-sample t test include: a null and alternative hypothesis, assumptions that must be checked, a test statistic t and its *p*-value, a decision, and a conclusion.

Hypothesis Testing Steps:

1. Determine appropriate null and alternative hypotheses about μ, and set α.
2. Check assumptions for performing the test.
3. Calculate the test statistic and determine the *p*-value.
4. Evaluate the p-value and report a conclusion in context of the problem.

The first step of a hypothesis test is to identify the hypotheses; this step is crucial, as it dictates the procedures for the remainder of the test. A one-sample *t*-test is used to test whether the mean of a quantitative variable is significantly different from some value. This value, the **test value** or μ_0, is given in the null hypothesis (H_0: $\mu = \mu_0$), and is often taken to be zero (i.e., H_0: $\mu = 0$). In Derek's test, his null hypothesis is H_0: $\mu = 800$.

The **alternative hypothesis**, H_a (or sometimes H_1), represents the experimenter's new model, or what the experimenter would like to show. The alternative hypothesis can take three different forms – it may be a denial of the null hypothesis (uses $\neq$; called a **two-sided test**), or it may specify a direction of interest (uses > or <; called a **one-sided test**). Notice that we never test for equality in H_a. Derek's alternative hypothesis will be H_a: $\mu > 800$.

The test relies on two key assumptions: (1) the data can be considered a **random sample** from the larger population of interest and (2) the data are observations from a **normally distributed <u>population</u>**. The Central Limit Theorem allows the assumption of normality to be more relaxed if our sample size, *n*, is large. (Generally, "large" means more than 30 observations, although it also depends somewhat on how serious the data depart from normality.) The sampling distribution applet that you worked with demonstrated the CLT using a *large* sample size of 25. Thus we will accept *at least 25* as large enough for inference about means.

The purpose of a significance test is to assess whether or not the observed data are consistent with the null hypothesis (within reasonable bounds of sampling variability). If the data seem unlikely to occur if the null hypothesis is assumed true, then we would reject the null hypothesis statement. To help us make this decision, we use a **test statistic**, which represents a summary of the data. Going back to the application, Derek samples 40 students at the University by asking about their monthly rent. From the sample he computes a sample mean rent of $900 as well as a standard error for this sample mean. His test statistic takes the form: $t = \frac{\bar{x}-\mu_0}{s/\sqrt{n}}$.

The **test statistic tells us how many standard errors the sample mean, $\bar{x}$, is from the test value, μ_0**. Derek calculated a test statistic of t = 1.2. His interpretation would state: The sample mean rent of $900 per month is 1.2 standard error away from the hypothesized mean rent of $800.

In hypothesis testing, another frequently reported value is the ***p*-value**, a number that is used to indicate the degree of significance of the data. The *p*-value is the probability of getting a test statistic as extreme or more extreme than the observed value of the test statistic, assuming the null hypothesis is true. In Derek's case, the *p*-value is the probability of observing a test statistic of 1.2 or larger, if the population mean rent in Ann Arbor really was 800 dollars more month. Note that more "extreme" means in the direction of Ha. His test statistic, t, has a t-distribution with n-1 degrees of freedom (df), where n represents the sample size. In Derek's test, the t statistic has a t (39) distribution.

We must decide in advance how much evidence against H_0 we will require for rejection. This designated amount of evidence is called the level of significance, denoted by α (alpha). Common values of α are 0.01, 0.05, and 0.10. If the *p*-value is less than or equal to α, Derek would make the decision to reject H_0. If Derek makes the decision to reject the null hypothesis, the results of the test are said to be statistically significant at level α. **A "significant" result in the statistical sense does not necessarily imply an "important" result in the practical sense.** It simply means that such a difference from the null hypothesis is not likely to happen just by chance.

Formula Card:

Population Mean	
Parameter	μ
Statistic	$\bar{x}$
Standard Error $\text{s.e.}(\bar{x}) = \frac{s}{\sqrt{n}}$	
Confidence Interval $\bar{x} \pm t^{*}\text{s.e.}(\bar{x})$	$\text{df} = n - 1$
Paired Confidence Interval $\bar{d} \pm t^{*}\text{s.e.}(\bar{d})$	$\text{df} = n - 1$
One-Sample *t*-Test $t = \frac{x - \mu_0}{\text{s.e.}(\bar{x})} = \frac{\bar{x} - \mu_0}{s/\sqrt{n}}$	$\text{df} = n - 1$
Paired *t*-Test $t = \frac{\bar{d} - 0}{\text{s.e.}(\bar{d})} = \frac{\bar{d}}{s_d/\sqrt{n}}$	$\text{df} = n - 1$

Warm-Up: Check Your Understanding

A researcher wants to determine if a filling machine at a cereal plant fills the boxes well on average or if the machine needs to be replaced. The machine is intended to fill the packages with 28 grams of cereal. He wants to test the machine to see if the average weight of cereal is less than the intended 28 grams. He takes a sample of 30 cereal boxes and measures the weight of each box. The histogram of the weights indicates some skewness.

Which result allows him to conduct the one sample t-test even though his histogram shows some skewness?

He performs a one sample t test and finds a statistically significant *p*-value.

If the significance level was 5%, what are possible options for the *p*-value?

0.0067 0.043 0.14 0.22

ILP: Is There Salmonella enteritidis in the Ice Cream?

Background: A massive multi-state outbreak of food-borne illness was attributed to *Salmonella enteritidis*. Epidemiologists determined that the source of the illness was ice cream. To determine the level of *Salmonella enteritidis* in the ice cream, they sampled nine production runs from the company that produced the ice cream. The levels of *Salmonella enteritidis* in Most Probable Number/Gram (MPN/g) are given in the **salmonella.sav** data set (Source Lyman Ott et. al., pg 232). The researchers would like to use these data to determine whether the average level of *Salmonella enteritidis* in the ice cream is greater than 0.3 MPN/g, since such levels are considered very dangerous using a 5% significance level. Time ordering of the data is present since the sample is collected from production runs.

Task: Perform a test to assess if the average level of *Salmonella enteritidis* in the ice cream is significantly larger than 0.3 MPN/g.

Clarify: Before conducting a test, here are a set of questions to ask yourself:

- What is the outcome or response variable?
- What type of variable is the response? *Categorical* *Quantitative*
- What type of parameter would be useful for summarizing this response?

Proportion *Mean*

Procedure: The appropriate inference procedure for this scenario is the one-sample t test and the specific parameter of interest is ________________ .

Hypothesis Test: You can complete the steps for conducting a test of hypotheses.

1. **State the Hypotheses:** H_0: __________ = __________ and H_a: __________ __________, where __________ represents:

   Determine Alpha: We were told the significance level was 5%.

Remember: Your hypotheses and parameter definition should always be a statement about the **population(s)** under study.

2. Checking the Assumptions

a. For this scenario, we need to assume that the data are a random sample. To check this assumption, we would make a __________ plot (if there was time order) of the observations, and look for __________________________ .

b. We also need to assume that the responses come from a normally distributed ________________________ . To check this assumption, we would make a ____________ plot.

c. Comment on each assumption below, using the provided graphs.

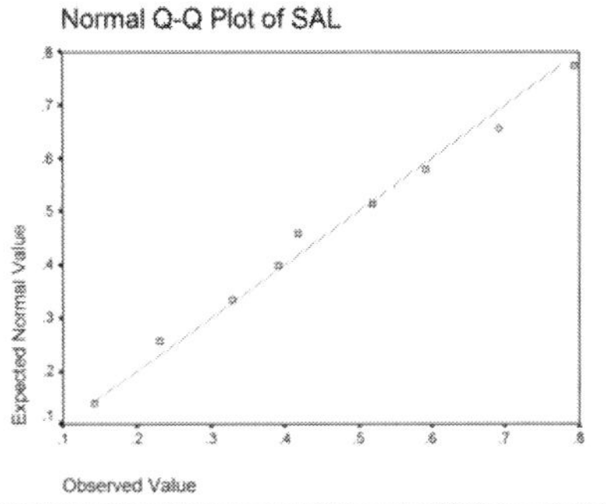

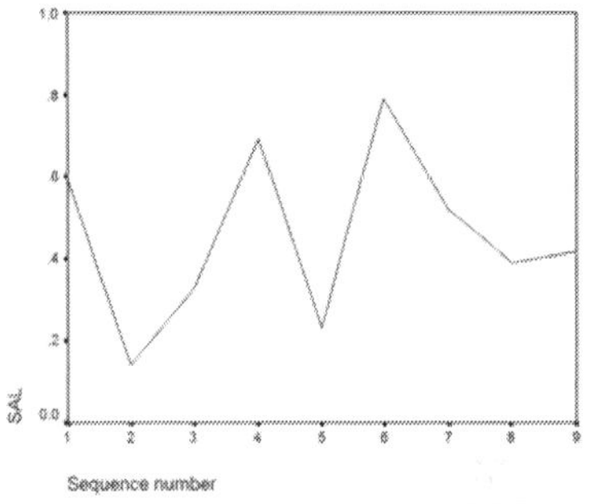

d. Is the assumption of normality for the population of ice cream really that important with a sample of size of 9 production runs? Why?

If the sample size were say 50 production runs instead of just 9, would this assumption be as important? Why?

3. Compute the Test-Statistic and Calculate the *p*-value

Test-Statistic

a. Generate the *t*-test output using **Analyze> Compare Means> One-Sample T Test**. Enter a test value of __________ (this is the value you are testing against from the null hypothesis).

b. What is the value of the test statistic? ____ = ___________

c. What is the distribution *of the test statistic* if the null hypothesis is true?

> **Note:** This is not the same as the distribution *of the population that the data were drawn from*, and will be the model used to find the *p*-value.

Calculate the *p*-value:

d. What is the SPSS reported two-tailed *p*-value? _____________ .

e. Report the correct *p*-value for this test. ______________.

f. Draw a picture of the *p*-value, with labels for the distribution and x-axis.

4. Evaluate the *p*-value and Conclusion

Evaluate the *p*-value:

What is your decision at a 5% significance level? **Reject H_0** **Fail to Reject H_0**

> **Remember:** Reject H_0 ⇔ Results statistically significant
> Fail to Reject H_0 ⇔ Results not statistically significant

Conclusion:

What is your conclusion *in the context of the problem*?

> **Note:** Conclusions should always include a reference to the population parameter of interest. Conclusions should not be too strong; you can say that you have sufficient evidence, but do NOT say that we have <u>proven</u> anything true or false.

ILP: Testing the Population Mean with *p*-Value Practice

For each of the following sets of hypotheses and test statistic values, provide a sketch of the *p*-value. You can assume the degrees of freedom are 20 throughout. Then use your computer with the `pval()` function in R (accessible from the Extra Review link on your course site), your calculator (if it has the capabilities), or the partial *t* distribution table provided below, and compute the *p*-value (or bounds for the *p*-value). Finally consider H_0 at a 5% significance level, and circle the corresponding statistical decision.

	Absolute Value of *t*-Statistic							
df	1.28	1.50	1.65	1.80	2.00	2.33	2.58	3.00
20	0.108	0.075	0.057	0.043	0.030	0.015	0.009	0.004

1. H_0: $\mu = 6$, H_a: $\mu > 6$, $t = 2.12$ **Reject H_0** **Do Not Reject H_0**
   Sketch:

2. H_0: $\mu = 0$, H_a: $\mu < 0$, $t = -3.20$ **Reject H_0** **Do Not Reject H_0**
   Sketch:

3. H_0: $\mu = 6$, H_a: $\mu \neq 6$, $t = -1.87$ **Reject H_0** **Do Not Reject H_0**
   Sketch:

Cool-Down: Check Your Understanding

Provide an **interpretation of the test statistic value** that you found in lab to test that the mean Salmonella level in ice cream was greater than 0.3 MPN/g.

The 95% confidence interval estimate for the population mean salmonella level in ice cream was found to be (0.2926 MPN/g, 0.6203 MPN/g). How would you explain the meaning of the 95% confidence level used in making the confidence interval?

Example Exam Question on One-Sample *t*-Tests

The Healthy Life Company produces canned whole pineapples. The production line is supposed to yield cans with an average weight of 1 kg. In order to assess that the production line is operating properly, the quality control statistician at Healthy Life selects a random sample of 36 cans produced over the course of the shift (keeping track of the order number). The latest sample gave a mean weight of 980 grams (1 kg = 1000 grams). Based on this sample, an estimate of the average distance that possible sample means are from the true population mean is about 8.33 grams.

a. The value of 8.33 grams described above corresponds to what statistical term or notation? (Be specific.)

b. The hypotheses to be tested are H_0: $\mu = 1000$ grams versus H_a: $\mu \neq 1000$ grams. To examine the data, various graphs were made. One graph is given below. State what assumption can be checked using the graph and comment on the validity of the assumption.

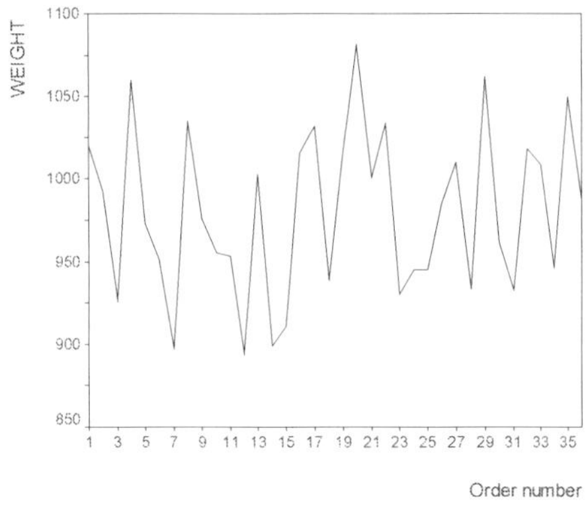

Assumption: ____.

Based on this graph, this assumption appears (circle one): ***valid*** ***not valid*** because:

c. The test is to be performed using a 5% significance level. The observed t-test statistic is $t = -2.4$ and the corresponding 2-tailed p-value is 0.022. Consider the statements below and clearly circle all those that are correct statements for this statistical analysis.

- *The null hypothesis is rejected at the 5% level.*
- *The results are not statistically significant at the 5% level.*
- *The probability that the null hypothesis is true is 0.022.*
- *If this process were operating as it should and repeated random samples of 36 cans were obtained, we would observe a t statistic of −2.4 or smaller or a t statistic of 2.4 or larger in about 2.2% of the repetitions.*

d. What would have been the p-value ...

i. if H_a: $\mu < 1000$ grams? ii. if H_a: $\mu > 1000$ grams?

e. If you were to compute a 95% confidence interval for the population mean weight, would you expect your interval to contain the value of 1000 grams? Why or why not?

Lab 8: Paired *t*-Test Procedures

Objective: In this lab, you will learn how to perform a hypothesis test in the case when we have two quantitative variables collected in pairs, called a **paired *t*-test**. You will make a confidence interval for and test hypotheses about the population mean difference, μ_d. Using these, you will be able to provide a statement about how confident you are regarding your interval estimate or in your decision.

Application: Mackenzie believes that college students can run a mile faster in the afternoon then they can in the morning. She has ten of her friends run a mile early in the morning and also late in the afternoon and she records their time in seconds. For each of her ten friends Mackenzie computes a difference: time to run the morning mile (in seconds) – time to run the afternoon mile (in seconds).

Runner	AM	PM	Difference (AM – PM)
Runner 1	633.8	618.9	14.9
Runner 2	588.9	569.6	19.3
Runner 3	619.4	630.9	-11.5
Runner 4	640.9	628.5	12.4
Runner 5	613.9	574.2	39.7
Runner 6	590.0	627.3	-37.3
Runner 7	613.4	603.3	10.1
Runner 8	568.6	593.1	-24.5
Runner 9	637.0	596.2	40.8
Runner 10	648.6	613.4	35.2

Overview: Matched or paired data results from a deliberate experimental design scheme. Mackenzie's scenario is one example of a paired design. Another example of a paired design is an experiment where rats are matched by weight, where one rat in each match receives a new diet and the other rat in the match receives a control diet. These types of design are called **paired data designs**. Note that paired designs can occur when you have two measurements on the same individual OR when you have two individuals that have been matched or paired prior to administering a treatment.

The inference procedures for a paired data design are based on the one-sample *t*-test procedures from the previous lab. The change is that we want to estimate or testing hypotheses about the population mean difference μ_d, which is generally compared to a hypothesized value of zero, indicating no difference on average.

The assumptions are similar to those assumptions made for the one sample t-test. Going back to the application, Mackenzie now must assume the sample of the ten differences in running time is a random sample. This can be verified by creating a time plot of those ten sample differences and checking for stability. She must also assume that the population of differences is normally distributed. Mackenzie can check this assumption by creating a QQ plot of those ten sample differences.

The test statistic t is created by taking the sample mean difference and dividing by the standard error of the sample mean difference, $t = \frac{\bar{d}}{se(\bar{d})}$. If the null hypothesis is true, this test statistic has a t distribution with n-1 degrees of freedom, where n is the number of pairs. For Mackenzie's 10 runners, the degrees of freedom will be 10 – 1 = 9.

Formula Card:

Population Mean		
Parameter	μ	
Statistic	$\bar{x}$	
Standard Error $\text{s.e.}(\bar{x}) = \frac{s}{\sqrt{n}}$		
Confidence Interval $\bar{x} \pm t^* \text{s.e.}(\bar{x})$		$\text{df} = n - 1$
Paired Confidence Interval $\bar{d} \pm t^* \text{s.e.}(\bar{d})$		$\text{df} = n - 1$
One-Sample t-Test $t = \frac{\bar{x} - \mu_0}{\text{s.e.}(\bar{x})} = \frac{\bar{x} - \mu_0}{s/\sqrt{n}}$		$\text{df} = n - 1$
Paired t-Test $t = \frac{\bar{d} - 0}{\text{s.e.}(\bar{d})} = \frac{\bar{d}}{s_d/\sqrt{n}}$		$\text{df} = n - 1$

Warm-Up: Check Your Understanding

Eight cars were run to determine their mileage, in miles per gallon (MPG). Then each car was given a tune-up, and run again to measure the mileage a second time. The difference in mileage was computed as difference = After MPG minus Before MPG. Assume that the selection of 8 cars represents a random sample of cars.

1. Select the appropriate alternative hypothesis to assess if on average mileage significantly improves after a tune up.

    a. $H_a: \mu_d > 0$

    b. $H_a: \mu_d < 0$

    c. $H_a: \mu_d \neq 0$

2. The researcher of course is hoping that the results of the experiment are statistically significant. What type of p-value would the researcher want to obtain? Select your answer.

    a. A large p-value.

    b. A small p-value.

    c. The magnitude of a p-value has no impact on statistical significance

ILP: Do Books Purchased from Barnes and Noble (In-store) Cost More on Average than If Purchased at Amazon.com (Online)?

Background: In recent years, the popularity of purchasing books online has increased dramatically, and the conventional bookstore no longer dominates the sales of books. The most influential factor that sways customers into purchasing online is lower prices. A group of Statistics 250 students decided to perform a comparison of the Amazon.com prices versus local Barnes and Noble bookstore prices based on a sample of 40 books, selected from a wide range of categories. For Amazon, a standard ground shipping of $4.29 and local state tax of 6% were included in the cost. The corresponding costs are available in the SPSS data set called **books.sav** (Source: Statistics group project, 2004). Do the data provide sufficient evidence to conclude that, on average, *Barnes and Noble (in-store) books are more expensive than Amazon.com books?*

Task: Perform a test regarding the mean difference in book price, μ_D, where the differences are computed as "*Barnes and Noble less Amazon*" (i.e. 'price at Barnes and Noble' minus 'price on Amazon').

Procedure: The appropriate inference procedure for this scenario is a paired t-test, and the specific parameter of interest is ___________________ .
NOTE: Why is this a paired procedure?

Hypothesis Test:

1. **State the Hypotheses:** H_0: _______________ and H_a: _______________ ,

   where the parameter__________ represents:

   Determine Alpha: We were told to use a significance level of 5%.

   Remember: Your hypotheses and parameter definition should always be a statement about the **population(s)** under study.

2. **Checking the Assumptions**
   a. For this scenario, we need to assume that the sampled differences are a random sample. Since the data was collected at one time point, a time plot is not appropriate to make. However we would want to learn more about how the books were selected.

   b. We also need to assume that the population of differences is normally distributed. To check this assumption, we would make a __________ plot of the ___________________. (What you would hope to see?)

3. Compute the Test-Statistic and Calculate the *p*-value

Test-Statistic

a. Generate the paired *t*-test output using **Analyze> Compare Means> Paired-Samples T Test.**

b. What is the value of the test statistic? ____ = ____________

c. Our sample mean difference is __________ standard errors below the hypothesized mean difference of zero.

d. What is the distribution *of the test statistic* if the null hypothesis is true?

> **Note:** This is not the same as the distribution *of the population that the data were drawn from*, and will be the model used to find the *p*-value.

Calculate the *p*-value:

e. What is the SPSS reported two-tailed *p*-value? _____________ .

f. What is the correct p-value for this test? ___________.

g. Draw a picture of the *p*-value, with labels for the distribution and x-axis.

4. Evaluate the *p*-value and Conclusion

Evaluate the *p*-value:

What is your decision at a 5% significance level? **Reject H_0** **Fail to Reject H_0**

> **Remember:** Reject H_0 ⇔ Results statistically significant
> Fail to Reject H_0 ⇔ Results not statistically significant

Conclusion:

What is your conclusion *in the context of the problem*?

> **Note:** Conclusions should always include a reference to the population parameter of interest. Conclusions should not be too strong; you can say that you have sufficient evidence, but do NOT say that we have <u>*proven*</u> anything true or false.

Cool-Down: Setting Up Hypotheses and Writing Conclusions

The Major League Baseball Organization would like to assess whether or not players have the same batting averages during the nighttime games and the daytime games at a 5% significance level. Eight players are selected at random and their nighttime and daytime batting averages are collected for a given period of time.

Let μ_d represent the population mean difference between the nighttime minus the daytime batting average. Write out the corresponding null and alternative hypothesis using the appropriate notation

Hypothesis Test: H_0:____________________

H_a:____________________

Why is this a paired t-test?

The *p*-value for this paired t-test was 0.028. Make a decision for this test and write a conclusion in context of the problem.

Example Exam Question on Paired *t*-Tests

A study was conducted to see how often two rooms of a sports facility were being used during the lunch hour. The number of people in each room was counted at 12:30 PM each Monday for 10 weeks. The results are summarized below.

Week	1	2	3	4	5	6	7	8	9	10
1 = Dance Studio	16	20	7	12	11	19	23	12	25	16
2 = Weight Room	11	13	8	14	12	15	9	15	18	10
d = Difference	5	7	-1	-2	-1	4	14	-3	7	6

Suppose you want to test the hypothesis of no difference between the utilization of the two rooms against the alternative that the dance studio is used by more people during the lunch hour on average. You enter the above data into SPSS, and conduct a paired *t*-test to obtain the output shown next.

Paired Samples Test

		Paired Differences							
					95% Confidence Interval of the Difference				
		Mean	Std. Deviation	Std. Error Mean	Lower	Upper	t	df	Sig. (2-tailed)
Pair 1	# using dance studio - # using weight room	3.60	5.34	1.69	-.22	7.42	2.13	9	.062

a. The observed test statistic is given as $t = 2.13$. State what this value tells you about the location of the sample mean difference of 3.6.

b. State the appropriate null and alternative hypotheses.

H_0: ____________________ H_a: ____________________

where __________ represents:

c. Report the p-value for the test in (b) and decision using a 10% significance level.

p-value: ____________________

Decision: (circle) ***Reject*** H_0 ***Fail to Reject*** H_0

d. A decision was made in part (c). Which type of error could have been made?

e. Which statements are **assumptions** required for performing the paired t-test?

- The population standard deviation for the difference in room use is known.
- The numbers using the dance studio are normally distributed.
- The difference in room use is normally distributed.
- The standard deviation for the number using the dance studio is equal to the standard deviation for the number using the weight room.
- The numbers using the dance studio are independent of the numbers using the weight room.

Lab 9: Two Independent Samples *t*-Test Procedure

Objective: In this lab, you will learn an important statistical technique that will allow you to compare two populations with respect to their means by looking at $\mu_1-\mu_2$. The aim is to help you understand the ideas behind confidence intervals, tests of significance, and the statistical language involved in the comparison of two population means.

Application: Jim believes that the mean GPA for all English majors is different from the mean GPA for all Math majors. He takes a random sample of students in the Math and English buildings and he ends up with 32 English majors and 35 Math majors in his sample. He has more Math majors, which leaves Jim with no method to pair an English major with Math major and no way to perform the paired t test.

Overview: The two independent samples *t* procedures (sometimes called Student's *t* procedures) are used when you want to compare the means of two populations that are not related or matched in any way. The idea is that we use the two sample means to estimate the corresponding population means. We may want to construct a confidence interval to estimate the difference between the two population means. Or, we may wish to test if the difference in the two population means equals a value specified in a null hypothesis; generally, the hypothesized test value is 0.

Going back to the application, we will assume that we have a random sample from each of the two populations. This means Jim will assume that he has a random sample of English majors and a random sample of Math majors. To check this assumption, he will create a time plot with the GPAs for the English majors and assess the stability as well as creating a separate time plot with the GPAs for the Math majors. The second assumption needed to perform the two independent sample t test is that the variable being measured has a normal model for each population, although possibly with different means. To assess this assumption, Jim will create two separate QQ plots; one of the GPAs for English majors and one of the GPAs for Math majors.

In addition, the two random samples are assumed independent of each other. However, there are two versions of independent samples *t* procedures: pooled and unpooled (also known as general). In a pooled *t* procedure, we assume equal population variances for the two populations of responses. Whether or not we can assume the two populations have equal variances will dictate which of the two procedures we use. Details on determining which version to use are below.

Checking the Equal Population Variance Assumption: There are several ways to check the equal population variances assumption required for a pooled *t*-test. Which one you will use will often depend on the information provided.

- ❑ **Side-by-side Boxplots:** Examine the IQRs of the sample data. If they are similar, then the assumption is valid. (NOTE: "Similar" in this sense does not mean that the boxes need to line up right next to each other. It means that the lengths or sizes of the boxes should be similar.) If one IQR is twice as large as the other, the assumption of equal population variances is not valid.
- ❑ **Sample Standard Deviations:** Since variance is standard deviation squared, if the sample standard deviations are similar, then the assumption is valid. If one standard deviation is twice as large as the other, the assumption of equal population variances doesn't hold.
- ❑ **Levene's Test for Homogeneity of Variances:** Levene's test is a hypothesis test, but it is a test about population variances rather than means. In Jim's case the null hypothesis is that the populations of English and Math majors do have equal variances (H_0: $\sigma_1^2 = \sigma_2^2$) and the alternative is that these populations have equal variances (H_a: $\sigma_1^2 \neq \sigma_2^2$). When the null hypothesis from the Levene's test is **NOT** rejected, it would be reasonable for Jim to perform the pooled *t*-test for the two population mean GPAs. If the null hypothesis from Levene's test **IS** rejected, then the assumption of equal population variances is unreasonable, and Jim should use the unpooled (general) version instead of the pooled. To make this determination, we will use an alpha of 10% for Levene's tests.

Common Population Standard Deviation: If Jim concludes that the populations of Math and English majors have the same variances, he is also stating that these populations have the same standard deviation. In order to run the pooled version of this test, he must estimate the common population standard deviation with the common sample standard deviation. This common standard deviation is written as s_p and Jim will calculate it by taking a weighted average of the standard deviation of English majors' GPAs and the standard deviation of Math majors' GPAs.

$$s_p = \sqrt{\frac{(n_1 - 1)s_1^2 + (n_2 - 1)s_2^2}{n_1 + n_2 - 2}}$$

Formula Card:

Two Population Means	
General	Pooled
Parameter $\mu_1 - \mu_2$	**Parameter** $\mu_1 - \mu_2$
Statistic $\bar{x}_1 - \bar{x}_2$	**Statistic** $\bar{x}_1 - \bar{x}_2$
Standard Error $\text{s.e.}(\bar{x}_1 - \bar{x}_2) = \sqrt{\frac{s_1^2}{n_1} + \frac{s_2^2}{n_2}}$	**Standard Error** pooled $\text{s.e.}(\bar{x}_1 - \bar{x}_2) = s_p\sqrt{\frac{1}{n_1} + \frac{1}{n_2}}$ where $s_p = \sqrt{\frac{(n_1 - 1)s_1^2 + (n_2 - 1)s_2^2}{n_1 + n_2 - 2}}$
Confidence Interval $(\bar{x}_1 - \bar{x}_2) \pm t^*(\text{s.e.}(\bar{x}_1 - \bar{x}_2))$ df = $\min(n_1 - 1, n_2 - 1)$	**Confidence Interval** $(\bar{x}_1 - \bar{x}_2) \pm t^*(\text{pooled s.e.}(\bar{x}_1 - \bar{x}_2))$ df = $n_1 + n_2 - 2$
Two-Sample *t*-Test $t = \frac{\bar{x}_1 - \bar{x}_2 - 0}{\text{s.e.}(\bar{x}_1 - \bar{x}_2)} = \frac{\bar{x}_1 - \bar{x}_2}{\sqrt{\frac{s_1^2}{n_1} + \frac{s_2^2}{n_2}}}$ df = $\min(n_1 - 1, n_2 - 1)$	**Pooled Two-Sample *t*-Test** $t = \frac{\bar{x}_1 - \bar{x}_2 - 0}{\text{pooled s.e.}(\bar{x}_1 - \bar{x}_2)} = \frac{\bar{x}_1 - \bar{x}_2}{s_p\sqrt{\frac{1}{n_1} + \frac{1}{n_2}}}$ df = $n_1 + n_2 - 2$

Warm-Up: Check Your Understanding

A student researcher, Jackie, is interested in learning about how much time students spend studying per week. She is curious to find out if study time is higher for female students (group 1) versus male students (group 2). Data from a survey of 47 female and 45 male undergraduate students is collected. One question asked students to report the number of hours that they study in a typical week.

A pooled independent t-Test was run and the difference in the two sample means, $\bar{x}_1 - \bar{x}_2$ has been computed. This difference is 3 standard deviations above the presumed difference of 0 for $\mu_1 - \mu_2$.

State the hypotheses to be tested and provide the test statistic value.

H_0: ______________________ versus H_a: ______________________

Test Statistic Value: ____________________

Which of the following is a correct graph of the p-value?

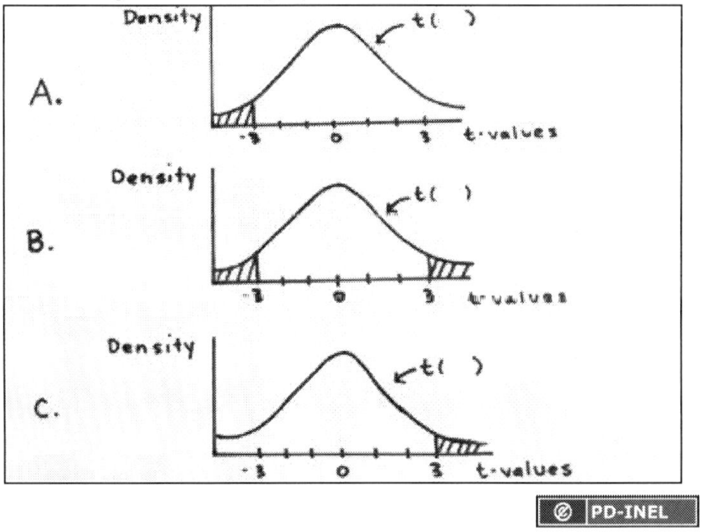

If the *p*-value for this test was 0.076,
are the results significant at a 5% level? **Yes** **No**

What are the appropriate degrees of freedom for the pooled t test? **91** **43** **90**

ILP: Do Men and Women Differ in Their SSHA Scores?

Background: A total of 38 college freshmen at a private college were administered the Survey of Study Habits and Attitudes (SSHA), a psychological test designed to measure motivation and attitude towards study habits in college students. The sampled students were a simple random sample consisting of 18 females and 20 males. Scores on the test range from a low of 0 to a high of 200 and it is known that they may explain collegiate success. School administrators are interested in whether or not there is a difference between the mean scores for males and females. The scores for females (Group 1) and males (Group 2) are listed in the **SSHA.sav** data set. (Source: Moore and McCabe (1999), pg 563).

***Task*:** Perform a test to assess if there is a difference between the mean score for women on the SSHA and the mean score for men on the SSHA.

Procedure: Since the SSHA score is a quantitative response and we have two sets of scores to compare, the appropriate inference procedure for this scenario is the two independent samples t-test, and the specific parameter of interest is ___________.
Note: Why is this not a paired procedure?

Hypothesis Test:.

1. **State the Hypotheses:** H_0: ______________ versus H_a: ________________ ,

   where __________ represents:

   Determine Alpha: We were told to use a significance level of 5%.

> **Remember:** Your hypotheses and parameter definition should always be a statement about the **population(s)** under study.

2. **Checking the Assumptions**

   a. Based on the information provided in the background of the problem, we will assume that we have independent, random samples of SSHA scores. One remaining assumption that needs to be checked is that each sample comes from a normally distributed population. The QQ plots for each sample are provided next.

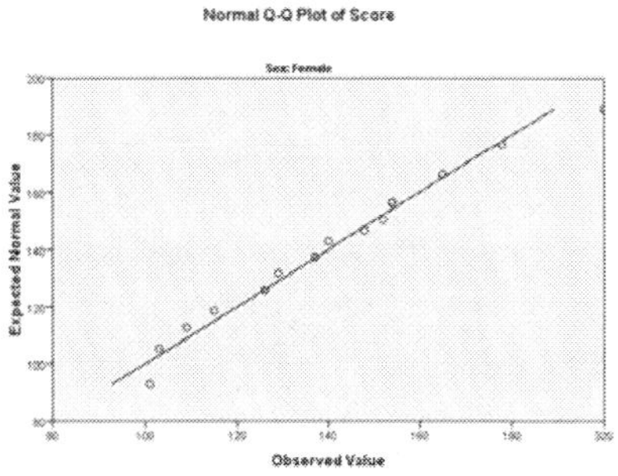

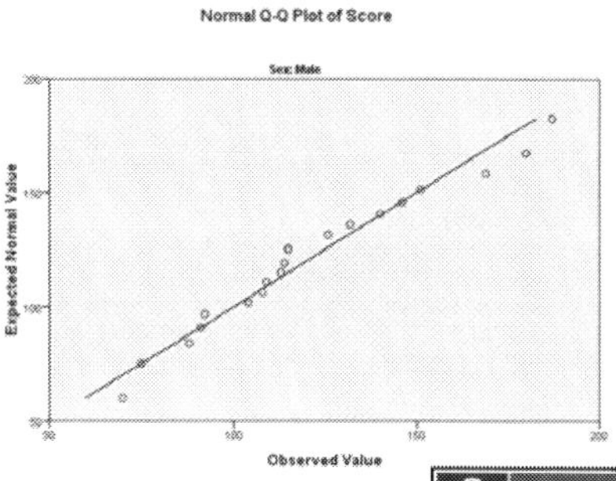

b. Does it appear that the assumption that each sample comes from a normally distributed population is met? Why?

c. What remaining assumption needs to be checked before we can conduct the two independent samples t-test?

Note: This assumption will be considered after the *t*-test output is generated. The *t*-test output will include the results from Levene's test that allows us to decide between the pooled and general two independent samples *t* procedure.

3. Compute the Test-Statistic and Calculate the *p*-value

Test-Statistic:

a. Generate the *t*-test output using **Analyze> Compare Means> Independent-Samples T Test**.

b. The output provides both the pooled (equal variances assumed) and unpooled (equal variances not assumed) versions of the test. Use the output as well as the provided side-by-side boxplots to determine which version of the test to use utilizing all three methods discussed earlier.

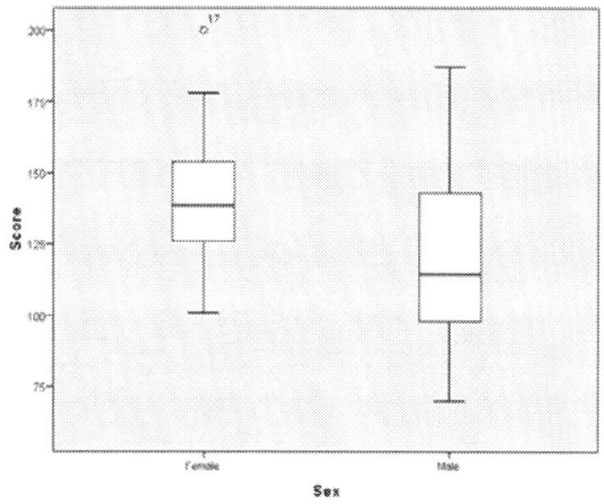

i. *Side-by-side Boxplots:* The side-by-side boxplots show that the IQRs are:

similar **not similar**

ii. *Sample Standard Deviations:* The standard deviation for females is ___________ and ___________ for males. This indicates that the standard deviations are: **similar** **not similar**

iii. *Levene's Test:* Write the appropriate hypotheses for Levene's test using correct notation:

H_0: _______________ H_a: ______________________

The Levene's test statistic is F = _______, with a *p*-value of _________. Therefore, we **can** **cannot** reject the null hypothesis that the population variances are equal.

Based on these results, we can say the assumption of equal population variances **is** **is not** valid. The symbol for the estimate of the common population standard deviation is ____ .

Calculate its value:

> **A Note on Output:** Don't confuse the Levene's test *p*-value with the *p*-value from the two independent samples *t*-test. Levene's test's *p*-value is labeled "Sig." in the Levene's section of output. Note the first line of output corresponds to the pooled procedure, while the second line corresponds to using an unpooled procedure.

c. Based on the version of the test you decided to conduct (pooled or general) in part f, what is the value of the test statistic? Make sure you are looking at the correct line of output.

d. What is the distribution *of the test statistic* if the null hypothesis is true?

> **Note:** This is not the same as the distribution *of the population that the data were drawn from*, and will be the model used to find the *p*-value.

Calculate the *p*-value:

e. What is the SPSS reported two-tailed *p*-value? _____________ .

f. Report the correct *p*-value for this test. _____________ .

g. Draw a picture of the *p*-value, with labels for the distribution and x-axis.

h. Provide an interpretation of the *p*-value (see Supplement 6).

4. Evaluate the *p*-value and Conclusion

Evaluate the *p*-value:

What is your decision at a 5% significance level? **Reject H_0** **Fail to Reject H_0**

> **Remember:** Reject H_0 ⇔ Results statistically significant
> Fail to Reject H_0 ⇔ Results not statistically significant

Conclusion:
What is your conclusion *in the context of the problem*?

> **Note:** Conclusions should always include a reference to the population parameter of interest. Conclusions should not be too strong; you can say that you have sufficient evidence, but do NOT say that we have <u>*proven*</u> anything true or false.

5. **Confidence Interval (CI):**
   a. Provide the corresponding 95% confidence interval from the output for the difference in the two population mean scores.

   b. Based on the confidence interval, would you reject the null hypothesis of no difference in population means at a 5% significance level?

   Circle one: **Yes** **No**
   Explain.

   Did your conclusion here match the one you made in the previous part 4?
   Circle one: **Yes** **No**

Cool-Down: Check Your Understanding

Consider the following sets of boxplots of scores between two age groups.

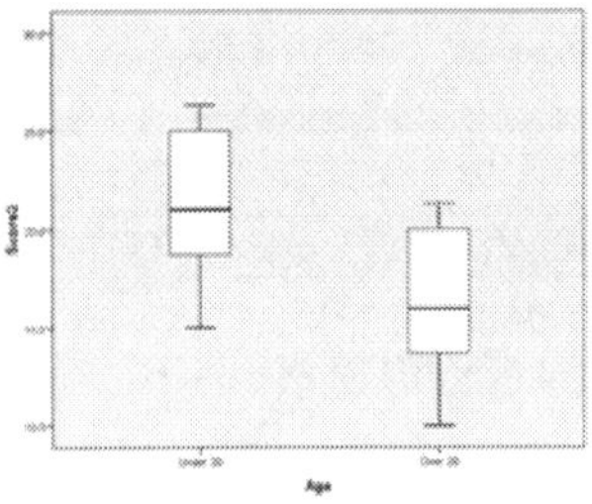

Set 1

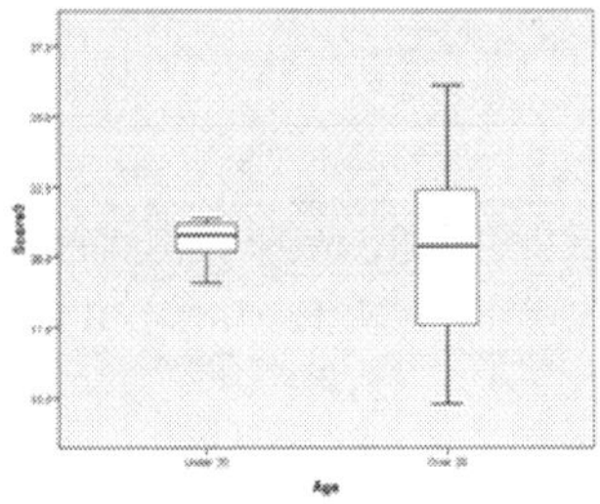

Set 2

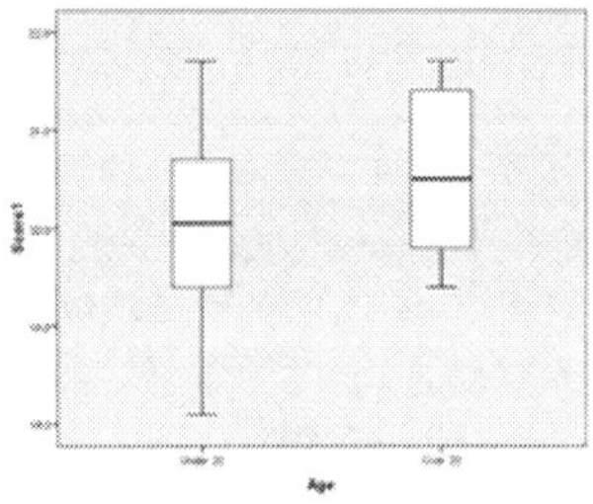

Set 3

a. Which sets that indicate a pooled test is appropriate:

Set 1 **Set 2** **Set 3**

b. For which sets are you most likely to reject the null hypothesis that the population mean scores are equal?

Set 1 **Set 2** **Set 3**

Example Exam Question on Two-Sample *t*-Test

A sample of 29 wines from a large distributor have been randomly selected for a wine tasting event. Three responses that are commonly measured, each on a quantitative scale, are aroma, flavor, and quality. The 29 wines were also classified by the region where each was produced. You are asked to provide a 95% confidence interval estimate for the difference in the population mean flavor rating for the two regions. SPSS output is provided. Assume all conditions for a two-independent samples t interval are met.

Group Statistics

	Region	N	Mean	Std. Deviation	Std. Error Mean
Flavor	1	17	4.376	.880	.213
	2	12	5.617	.915	.264

Independent Samples Test

		Levene's Test for Equality of Variances		t-test for Equality of Means						
									95% Confidence Interval of the Difference	
		F	Sig.	t	df	Sig. (2-tailed)	Mean Difference	Std. Error Difference	Lower	Upper
Flavor	Equal variances assumed	.028	.868	-3.68	27	.0010	-1.240	.337	-1.932	-.548
	Equal variances not assumed			-3.65	23.236	.0013	-1.240	.340	-1.942	-.538

a. Based on the output, give two reasons why you should report the ***pooled*** version of the two independent samples t interval for these data. Be specific, that is, **include values** in each of your explanations.

Because ...

Because ...

b. Which of the following values must be the pooled estimate of the common population standard deviation?

Circle one: 0.028 0.868 0.337 0.894 0.799

c. The pooled 95% confidence interval for the difference in the population mean flavor rating for region 1 versus region 2 is given by (-1.932, -0.548).

The probability that $(\mu_1-\mu_2)$ lies between –1.932 and –0.548 equals 0.95.

True *False*

Lab 10: One-Way Analysis of Variance (ANOVA)

Objective: In this lab you will perform a **one-way analysis of variance**, often abbreviated **ANOVA**. We have already seen that the two independent samples *t*-test can be used to compare the means of two populations (when the samples are independent). However, when we want to compare the means of *three or more populations*, we turn to ANOVA. You can think of ANOVA as an extension of the two independent sample pooled *t*-test since it compares several population means and **requires** the assumption that the populations have equal variances.

Application: Andy is conducting a test for UM Hospital to explore the effects of a new antibiotic drug. They wish to explore if this drug has the same effects on the mean white blood cell count for the four different age populations*:* 1 = 0 to 19 years, 2 = 20 to 29 years, 3 = 31 to 40 years, and 4 = 40 years and older. They need a test that would allow them to compare the mean white blood cell count for these four populations. In this example, the number of populations under study is *k*=4, and the total sample size is 24 (6 people in each age group).

Overview: ANOVA is a statistical tool for analyzing how the mean value of a quantitative response (or dependent) variable is affected by one or more categorical variables, known as treatment variables or factors. While ANOVA allows us to compare the means of more than two populations, it can only tell us whether differences appear to exist, not specifically which population means are different. Consequently, the appropriate hypotheses for ANOVA are H_0: $\mu_1 = \mu_2 = \ldots = \mu_k$ (that the population mean responses are equal, where k is the number of populations or treatment groups) and H_a: at least one of the population mean responses, μ_i, is different. Going back to the application, for Andy's test H_0: $\mu_1 = \mu_2 = \mu_3 = \mu_4$ and the H_a: at least one of the population mean white blood cell counts, μ_i, is different.

As with our other hypothesis tests, several assumptions are required for ANOVA. Andy will need to assume that the white blood cell count measurements in each of the four age populations follows a normal model. To check this assumption, he can create four QQ plots, one for each sample of measurements. He will also need to assume that the four populations of white blood cell counts have equal variance. The equal population variance assumption can be checked as done in the two independent samples t-test – side-by-side boxplots, comparing sample standard deviations, and Levene's test. Further, the data are assumed to consist of independent random samples.

The analysis of variance that Andy will run involves decomposing the total variation of the white blood cell count into two parts: (1) that due to variation among the four sample white blood cell counts (*between groups* variation) and (2) that due to the natural variation of the white blood cell count in each of the four age groups (variation due to error, or *within groups* variation): **SS Total = SS Groups + SS Error**.

If Andy finds that the sum of squares between groups (SS Groups) is large relative to the sum of squares within groups (SS Error), it implies that the model of different white blood cell means explains a significant portion of the observed variability.

In order to determine what is "relatively large," the sum of squares values are divided by their respective degrees of freedom, creating what are called mean square terms. The degrees of freedom for SS Groups is the number of treatment groups, k, minus one (k - 1). In Andy's test, the degrees of freedom for SS Groups is equal to3. For SS Error, the degrees of freedom is the total number of observations, N, minus the number of treatment groups (N - k). For Andy's ANOVA procedure, the degrees of freedom for the SS Error is equal to 20.

$$\text{MS Groups} = \text{SS Groups}/(k - 1)$$

$$\text{MS Error} = \text{SS Error}/(N - k)$$

The ratio of these two mean squares forms the ***F*-statistic**, which has numerator degrees of freedom (k - 1) and denominator degrees of freedom (N - k). Note in the following equation that MSE stands for MS Error.

$$F = \frac{\text{Variation among sample means}}{\text{Natural variation within groups}} = \frac{\text{MS Groups}}{\text{MSE}}$$

We can view this ***F*-statistic** as the ratio of two estimators of the common population variance, σ^2, where the denominator (MSE) is a good (unbiased) estimator, and the numerator (MS Groups) is only good when the H_0 is true (otherwise, it tends to overestimate σ^2). If Andy's data results in a large F value then there is some evidence against the null hypothesis of equal population white blood cell count means.

If Andy rejects the null hypothesis, indicating that at least one of the population mean white blood cell count is different, then we can turn to a **multiple comparisons procedure** for determining which population mean(s) appear to be different and how they differ. The most common method to analyze this is by looking at the set of all *pairwise comparisons*. Two equivalent techniques can be used by Andy for each pair of means: either perform a hypothesis test to see if two population mean white blood cell counts are significantly different, or construct a confidence interval for the difference in population mean white blood cell counts to see whether the value of 0 is contained in the interval. Specifically, a multiple comparisons procedure called **Tukey's procedure**, which is available in most computer packages, controls for the overall Type I error rate (overall significance level) or the overall confidence level.

Formula Card:

One-Way ANOVA		
SS Groups = SSG = $\sum_{\text{groups}} n_i(\bar{x}_i - \bar{x})^2$	MS Groups = MSG = $\dfrac{\text{SSG}}{k-1}$	**ANOVA Table**
SS Error = SSE = $\sum_{\text{groups}} (n_i - 1)s_i^2$	MS Error = MSE = $s_p^2 = \dfrac{\text{SSE}}{N-k}$	
SS Total = SSTO = $\sum_{\text{values}} (x_{ij} - \bar{x})^2$	$F = \dfrac{\text{MS Groups}}{\text{MS Error}}$	
Confidence Interval $\bar{x}_i \pm t^* \dfrac{s_p}{\sqrt{n_i}}$	df = $N - k$	Under H_0, the F statistic follows an $F(k-1, N-k)$ distribution.

Source	SS	DF	MS	F
Groups	SS Groups	$k-1$	MS Groups	F
Error	SS Error	$N-k$	MS Error	
Total	SSTO	$N-1$		

Warm-Up: Content of TV Shows

A study examined whether the content of TV shows influence the ability of viewers to recall brand names of items featured in the commercials. The researchers randomly assigned 90 adults to watch one of three programs (30 to each). One program had violent content, another sexual content, and the third neutral content. Each program contained the same nine commercials. After the shows ended, the subjects were asked to recall the brands of products that were advertised. The ANOVA table based on the data is provided below.

Number of Brands Recalled — **ANOVA**

	Sum of Squares	df	Mean Square	F	Sig.
Between Groups	11.8	2	5.90	7.47	0.001
Within Groups	68.6	87	0.79		
Total	80.4	89			

a. The researcher would like to assess if there is any effect of the program type on the average number of brands recalled, namely, test H_0: $\mu_1 = \mu_2 = \mu_3$. Provide the appropriate alternative hypothesis in the context of the problem.

H_a: ____

b. One assumption is the model for number of brands recalled for each population is normal. What graph(s) would you would make to assess this assumption?

c. At the 5% level, was there a significant effect of the program type on the average number of brands recalled? Explain your answer.

ILP: Is There a Difference Among the Mean Freshman GPAs for Three Different Socioeconomic Classes?

Background: Sociologists often conduct experiments to investigate the relationship between socioeconomic status and college performance. Socioeconomic status is generally partitioned into three classes: lower, middle, and upper. Consider the problem of comparing the mean grade point average (GPA) of college freshmen across the three socioeconomic populations. The GPAs for random samples of seven college freshmen from each of the three socioeconomic classes were selected from a university's files at the end of the first academic year. The data are in the **gpa.sav** data set. (Source: Mendenhall and Sincich, 1996, page 589)

Task: Perform a test to assess whether the mean freshman GPAs among the three socioeconomic classes differ. If there is sufficient evidence to indicate significant differences, determine which groups differ and how.

Procedure: We want to compare three populations with respect to a quantitative response (GPA). The appropriate inference procedure for this scenario is ANOVA and the value of *k* for this problem is ________ .

Hypothesis Test:

1. **State the Hypotheses:** H_0: ______________________________

   versus H_a: __________________ ,

   where ___________ represents:

   Determine Alpha: We were told to use a significance level of 5%.

   > **Remember:** Your hypotheses and parameter definition should always be a statement about the **population(s)** under study.

2. **Checking the Assumptions**
   We need to assume:
   - The *k* samples are _____________ from each other.
   - Each sample is a random sample. To check this assumption, we would make a _______________ plot (if there was time order) for each sample.
   - Each sample needs to come from a normally distributed population. To check this assumption, we would make a _______ plot for each sample.
   - All k populations have equal ____________________ .

a. Based on the description about how these samples were collected, can we assume we have random and independent samples?

Note that there is no time order for this data. If there were, since you need EACH sample to be a random sample, how many time plots would you need to make to check this assumption? _______ time plot(s)

b. Construct the Q-Q plots necessary to check the assumption about normally distributed populations. Recall the command **Data> Split File**. Does it appear that the assumption that each sample comes from a normally distributed population is met? Why?

> **Note:** The equal population variances assumption will be considered after the ANOVA output is generated.

3. Compute the Test-Statistic and Calculate the *p*-value

Test-Statistic

a. Generate the ANOVA output using **Analyze> Compare Means> One-Way ANOVA**. (Make sure you have turned off any split file features you may have been using in your assumptions checks.) Under **Options**, select the **Descriptive** (which gives you sample means and standard deviations) and the **Homogeneity of variance test** (Levene's test) options.

b. Choose a way to determine if the assumption of equal population variances is valid. Check the assumption and comment.

The assumption of equal population variances seems: **valid** **not valid**

The symbol for the estimate of the common population standard deviation is ____, which for this problem is found to be ____________ .

c. What is the value of the test statistic? ____ = ____________

d. What is the distribution *of the test statistic* if the null hypothesis is true?

Note: This is not the same as the distribution *of the population that the data were drawn from*, and will be the model used to find the *p*-value.

Calculate the *p*-Value:

e. What is the SPSS reported *p*-value? ____________

f. Draw a picture of the *p*-value, with labels for the distribution and x-axis. Use the *pval()* function in R to check your work.

4. Evaluate the *p*-value and Conclusion

Evaluate the *p*-value:

What is your decision at a 5% significance level? **Reject H_0** **Fail to Reject H_0**

Remember: Reject H_0 ⇔ Results statistically significant Fail to Reject H_0 ⇔ Results not statistically significant

Conclusion:

What is your conclusion *in the context of the problem*?

Note: Conclusions should always include a reference to the population parameter(s) of interest. They should not be too strong; you can say that you have sufficient evidence, but do NOT say that we have <u>proven</u> anything true or false.

5. **Follow-up Analyses:** If ANOVA has indicated that there appears to be significant differences between two or more of groups, we can use a multiple comparison test to tell us which groups appear to be different and by how much.

   a. Obtain the multiple comparisons output using **Analyze> Compare Means> One-Way ANOVA**. Under **Post Hoc...**, choose **Tukey**, which has a default significance level of 0.05. The multiple comparisons output contains both *p*-values and confidence intervals for every possible pairwise comparison of groups; either can be used to determine where differences exist. The *p*-values that are less than or equal to 0.05 or confidence intervals that do NOT contain 0 indicate a difference between those two population means.

   b. Summarize the findings about the differences in population means for the GPAs of freshmen in the different socioeconomic classes. Which pairs are significantly different at the 5% level?

   c. Calculate a 95% confidence interval for the mean GPA for the middle class group. The sample mean GPA for the 7 subjects in the group was 3.25.

Cool-Down: Check Your Understanding

Complete the following sentences by circling words or filling in blanks as necessary.

The *p*-value of 0.025 from this activity implies that if this study were repeated many times, we would see an *F* test statistic of 4.579 or **greater** **less** in about __________ % of repetitions if the population means were really all equal.

ANOVA procedures can be thought of as an extension of the two independent samples **pooled** **unpooled** *t*-test, and hence requires the assumption of equal **population** **sample** variances.

One way to check this assumption is to use Levene's test and see if the *p*-value is **greater than** **less than or equal to** 0.10 (or any reasonable significance level).

Example Exam Question on ANOVA

A study was conducted to compare the effects of two different therapy treatments and a control condition on weight gain in anorexic girls. Group 1 was the control condition subjects that received no intervention, Group 2 subjects received a cognitive-behavioral treatment condition, and Group 3 subjects received a family therapy condition. The response was weight gain over a fixed time period.

a. The ANOVA output provided below is used to test a set of hypotheses.

Gain in Weight

	Sum of Squares	df	Mean Square	F	Sig.
Between Groups	601.916	2	300.958	5.421	.007
Within Groups	3331.037	60	55.517		
Total	3932.953	62			

i. State the null and alternative hypotheses.

H_0: ______________________________

H_a: ______________________________

ii. The *p*-value for this test is reported as 0.007. Draw a sketch of the appropriate distribution showing how the *p*-value was determined for this ANOVA study. Provide all details.

b. Multiple comparisons were performed on the weight gain data.

Dependent Variable: Gain in Weight
Tukey HSD

(I) Condition	(J) Condition	Mean Difference (I-J)	95% Confidence Interval Lower Bound	95% Confidence Interval Upper Bound
Control	Cog Behav	-3.65	-8.77	1.48
	Family	-8.29	-14.36	-2.22
Cog Behav	Control	3.65	-1.48	8.77
	Family	-4.64	-10.58	1.30
Family	Control	8.29	2.22	14.36
	Cog Behav	4.64	-1.30	10.58

Circle all pairs that are significantly different (using a 5% significance level).

control versus cognitive behavior

control versus family therapy

cognitive behavior versus family therapy

Lab 11: Simple Linear Regression

Objective: In this lab, you will examine relationships between two quantitative variables using a graphical tool called a scatterplot. You will interpret scatterplots in terms of form, direction, and strength of the relationship, and use it to assess the appropriateness of using a simple linear regression model to describe the relationship between the two variables. If appropriate, you can then perform a simple linear regression analysis to produce an estimated model that can be used to predict the value of the response *y* for a given value of the predictor *x*. You will learn how to perform hypothesis tests and compute confidence intervals in regression; including learning how to check the assumptions needed for these inference procedures to be valid

Application: Pam believes that the number of Facebook friends a person has could interfere with the number of hours they spends socializing offline. She wants to be able to predict the number of hours a person spends socializing offline for a given number of Facebook friends. She will collect data from her friends and examine the relationship between these two variables. Pam will be able to fit a simple linear regression model if the relationship between these variables is linear.

Overview: A regression model describes how the mean of one variable is thought to depend on the value of one or more other variables. If we think the number of Facebook friends may explain changes in the amount of time spent socializing online, we call the number of Facebook friends an **explanatory** variable (or **predictor** variable or **independent** variable) and the amount of time spent socializing offline is called the **response** variable (or **dependent** variable).

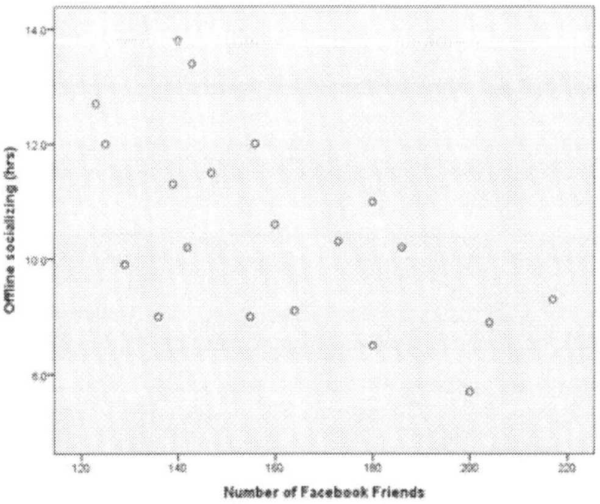

To start, we use a scatterplot to display the relationship between two quantitative variables, plotting the number of Facebook friends on the *x*-axis and the amount of time spend socializing offline on the *y*-axis. In examining the relationship, Pam looks at the overall pattern showing the **form**, which appears to be linear. She notes that the **direction** of the form is negative and **strength** of the relationship is moderate. She notes that there are no apparent **outliers**.

One of the many misconceptions about regression arises from the concept of association. Scatterplots can show the association between variables, but Pam should remember that *correlation does not imply causation*. For example: weekly flu medication sales and weekly sweater sales for an area with extreme seasons would exhibit a positive association because both tend to go up in winter and down in summer. However, neither causes the other. The observed association between two variables is sometimes due to other factors, such as confounding variables.

This correlation value, **r**, explains the strength of the linear relationship between x and y. The correlation can take on values between -1 and 1. The sign of the correlation also describes the direction of the linear relationship. The correlation between number of Facebook friends and time spent socializing offline is reported to be -0.59, which confirms what Pam observed in her scatterplot. The square of this correlation is known as the coefficient of determination and also has an important interpretation. Pam calculates this coefficient to be 0.348 and she can interpret this value by stating that 34.8% of the variation in the amount of time spent socializing online can be explained by the linear relationship between number of Facebook friends and time spent socializing online.

Since Pam's scatterplot suggested that the dependence of amount of time spent socializing online on the number of Facebook friends can be summarized by a straight line, the **least squares regression line** can be calculated. The least squares regression line is the line that minimizes the sum of the squared vertical distances of the data points to the line – hence the name **least squares**. This fitted line can be used to describe the linear relationship between the amount of time spent socializing offline and the number of Facebook friends and to predict the amount of time spent socializing online for a given number of Facebook friends.

The distances from the observed amount of time spend socializing offline to the predicted amount of time spent socializing offline are known as the **residuals**. These residuals are estimates of the true error terms associated with Pam's model.

Pam has fit a **simple linear regression model,** which is what will also be fit during the lab. Pam's regression is based on a linear model relating the number of Facebook friends to the mean amount of time spend socializing offline as follows: E(*time spent socializing offline*) = β_0 + β_1(*number of Facebook friends)*. Here β_0 and β_1 are parameters – fixed but unknown constants. Specifically, β_0 is the population *y*-intercept (the amount of time spent socializing offline when the number of Facebook friends is zero) and β_1 is the population slope (the change in the mean time spent socializing offline for every additional Facebook friend). These two values are unknown, but can be estimated using the least squares criterion. The resulting estimated regression line is generally written as: $\hat{y} = b_0 + b_1(x)$. The estimates, b_0 and b_1 are referred to as the **least squares estimates** of β_0 and β_1.

There are also several assumptions that Pam must check in order for inferences to be valid. First, the time spent socializing offline must have a normal distribution with a mean that varies linearly with the number of Facebook friends and a standard deviation that does not depend on the number of Facebook friends. To check her assumption Pam would create a residual plot: a plot with residuals on the y axis and the number of Facebook friends on the x axis. For this assumption to hold, the residuals be randomly scattered.

Pam must also assume that the error terms are normally distributed and are identically distributed. To check this assumption about the normal distribution she will create a QQ plot of the residuals. To check if the errors are identically distributed, she will create a time plot of the residuals and look for stability.

Formula Card:

Regression	
Linear Regression Model **Population Version:** Mean: $\mu_Y(x) = E(Y) = \beta_0 + \beta_1 x$ Individual: $y_i = \beta_0 + \beta_1 x_i + \varepsilon_i$ where ε_i is $N(0, \sigma)$ **Sample Version:** Mean: $\hat{y} = b_0 + b_1 x$ Individual: $y_i = b_0 + b_1 x_i + e_i$	**Standard Error of the Sample Slope** $\text{s.e.}(b_1) = \frac{s}{\sqrt{S_{XX}}} = \frac{s}{\sqrt{\sum (x-\bar{x})^2}}$ **Confidence Interval for β_1** $b_1 \pm t^* \text{s.e.}(b_1)$ df $= n-2$ **t-Test for β_1** To test $H_0: \beta_1 = 0$ $t = \frac{b_1 - 0}{\text{s.e.}(b_1)}$ df $= n-2$ or $F = \frac{MSREG}{MSE}$ df $= 1, n-2$
Parameter Estimators $b_1 = \frac{S_{XY}}{S_{XX}} = \frac{\sum (x-\bar{x})(y-\bar{y})}{\sum (x-\bar{x})^2} = \frac{\sum (x-\bar{x})y}{\sum (x-\bar{x})^2}$ $b_0 = \bar{y} - b_1 \bar{x}$	**Confidence Interval for the Mean Response** $\hat{y} \pm t^* \text{s.e.(fit)}$ df $= n-2$ where $\text{s.e.(fit)} = s\sqrt{\frac{1}{n} + \frac{(x-\bar{x})^2}{S_{XX}}}$
Residuals $e = y - \hat{y} =$ observed y − predicted y	**Prediction Interval for an Individual Response** $\hat{y} \pm t^* \text{s.e.(pred)}$ df $= n-2$ where $\text{s.e.(pred)} = \sqrt{s^2 + (\text{s.e.(fit)})^2}$
Correlation and its square $r = \frac{S_{XY}}{\sqrt{S_{XX} S_{YY}}}$ $r^2 = \frac{SSTO - SSE}{SSTO} = \frac{SSREG}{SSTO}$ where $SSTO = S_{YY} = \sum (y-\bar{y})^2$	**Standard Error of the Sample Intercept** $\text{s.e.}(b_0) = s\sqrt{\frac{1}{n} + \frac{\bar{x}^2}{S_{XX}}}$ **Confidence Interval for β_0** $b_0 \pm t^* \text{s.e.}(b_0)$ df $= n-2$
Estimate of σ $s = \sqrt{MSE} = \sqrt{\frac{SSE}{n-2}}$ where $SSE = \sum (y-\hat{y})^2 = \sum e^2$	**t-Test for β_0** To test $H_0: \beta_0 = 0$ $t = \frac{b_0 - 0}{\text{s.e.}(b_0)}$ df $= n-2$

Warm-Up: Check Your Understanding

For which scatterplot(s) does a linear regression analysis seem appropriate?

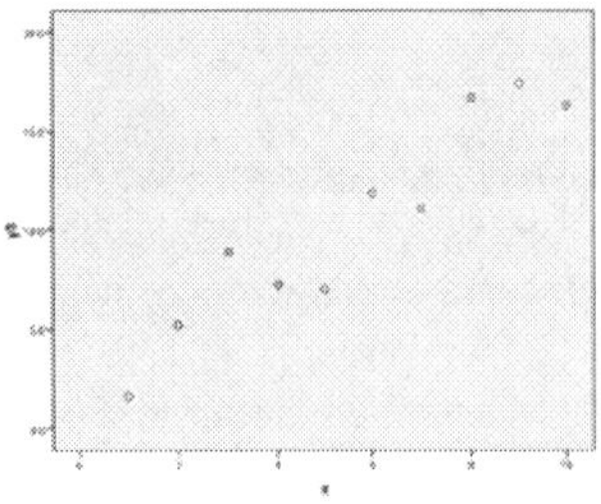

Plot 1

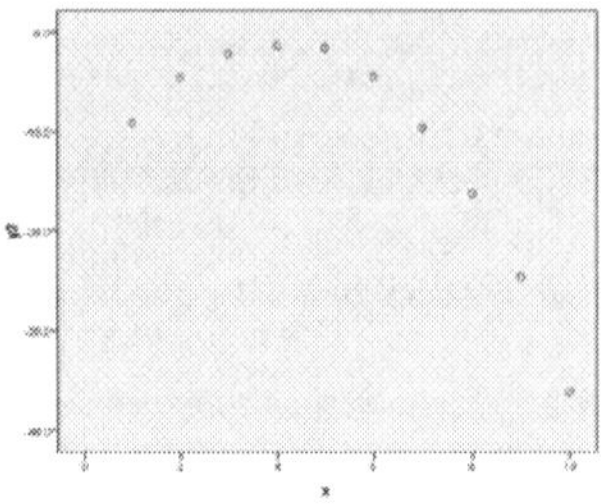

Plot 2

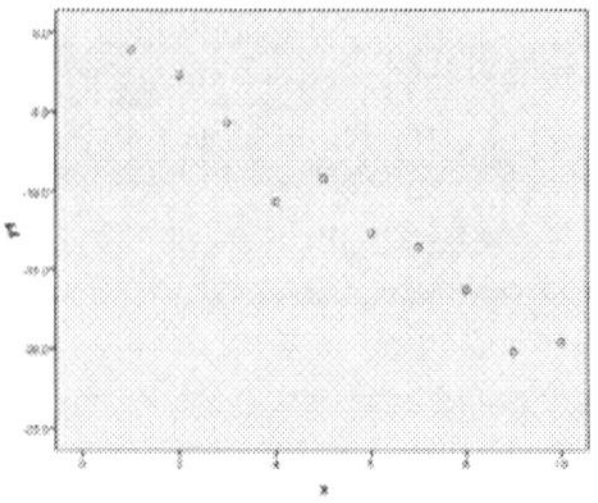

Plot 3

ILP: Using a Scatterplot to Display the Relationship

Background: The local library is interested in determining if the time spent watching TV is taking over a young person's reading time. They are interested in predicting the number of books read the past year from the number of minutes spent watching TV on a typical week night. The library took a sample of young patrons and recorded this information. The data set **TVandBooks.sav** contains the number of books a patron has read yearly (*Books*) as well as the number of minutes the patron spends watching television on a weekly basis (*TV*).

***Task*:** Is there a relationship between the amount of time spent watching television and the number of books a person has read? Produce a scatterplot to help examine the plausibility of a linear relationship between minutes spent watching television and number of books read. Note: You are interested in predicting the number of books read from the number of minutes spent watching TV.

Procedure: We have **two variables** being measured – the number of books read (which is **quantitative** and playing the role of the **response**) and the number of minutes spent watching TV (which is also **quantitative** and plays the role of the **explanatory** variable). The goal is to assess the **relationship** between these two quantitative variables. The appropriate inference procedure for this scenario is **simple linear regression**.

1. Open the data set and produce a scatterplot: **Graphs> Legacy Dialogs> Scatter/Dot**. Note that your explanatory variable should be on the x-axis (independent), and your response variable should be on the y-axis (dependent). Sketch the general pattern of the plot below. Make sure to label your axes.

2. Based on the scatterplot, does there appear to be a linear relationship between minutes of television watched and the number of books read? Explain.

3. Are there any unusual observations or outliers present?

4. Based on the scatterplot, which direction is the association between these two variables?

5. Does there appear to be a *strong* relationship between the minutes of television watched and the number of books read?

ILP: Describing a Linear Relationship with a Regression Line

If we are satisfied that a linear model seems to describe the relationship between the minutes of television watched and the number of books read, we are ready to estimate that model and use it to predict Books on the basis of TV.

Task: Fit a linear model to the data. If you have questions about the regression output after the activity, refer to Supplement 8 in this workbook for more details.

1. Obtain the linear regression output using **Analyze> Regression> Linear**. Make sure to enter the appropriate dependent and independent variables. Report the estimated regression line (predicting equation):

   ____

2. Interpret the estimated slope b_1 in terms of change in number of books read.

4. Report the coefficient of correlation, r, between Books and TV: r = __________

5. Report the coefficient of determination, r^2, and interpret it: r^2 = ____________

   Interpretation:

6. Use the regression line to predict the yearly number of books read for a person who watches 100 minutes of television weekly. How does it compare to the observed yearly number of books read for the patron who has watched 100 minutes of television weekly?

ILP: Is There a Significant (Non-Zero) Linear Relationship Between Teen Birth Rate and Poverty Rate?

Is the explanatory variable, time spent watching TV, a useful linear predictor for the number of books read? Is there a significant, non-zero linear relationship between time spent watching television (*TV*) and number of books read (*Books*)?

Remember that another way to make inferences about the significance of the linear relationship is through a confidence interval for the population slope. Further, recall the basic form of a confidence interval: point estimate ± (a few) standard errors.

Most standard computer regression output provides the slope estimate and its standard error, and the "few" will correspond to a t^* value for the corresponding confidence level with degrees of freedom for regression of $n - 2$. Since a confidence interval provides a range of reasonable values for the parameter, it can be used to perform two-sided hypothesis tests by seeing whether the hypothesized value falls in the interval or not.

Task: Assess if the explanatory variable TV is significant in the linear model.

Hypothesis Test:

1. **State the Hypotheses:** H_0: _______________ and H_a: ___________________ ,

   where ___________ represents:

   Determine Alpha: We were told to use a significance level of 5%.

> **Remember:** Your hypotheses and parameter definition should always be a statement about the **population(s)** under study.

2. **Checking the Assumptions**
   Assumptions: Covered in the next activity.

3. **Compute the Test-Statistic and calculate the *p*-value:**
   Test-Statistic
   a. Using the computer output, which two test statistics could you use to test these hypotheses? Give the value for each test statistic listed in part a.

   b. Which would not be appropriate for conducting a one-sided version of the alternative hypothesis?

Calculate the *p*-Value:

c. What is the reported *p*-value (for both test statistics)?________________

4. **Evaluate the *p*-value and Conclusion**

Evaluate the *p*-value:

What is your decision at a 5% significance level? **Reject H_0 Fail to Reject H_0**

Remember: Reject H_0	⇔ Results statistically significant
Fail to Reject H_0	⇔ Results not statistically significant

Conclusion:

What is your conclusion *in the context of the problem*?

Note: Conclusions should always include a reference to the population parameter(s) of interest. They should not be too strong; you can say that you have sufficient evidence, but do NOT say that we have *proven* anything true or false.

5. **Confidence Intervals (CI):**
   a. Generate the confidence interval for estimating the population slope using **Analyze> Regression> Linear**. Under **Statistics**, choose the **Confidence intervals**, which has a default level of 95%.

      Give the 95% confidence interval for the population slope.

   b. Provide an interpretation of the resulting interval in context.

   c. Based on the confidence interval, would you reject the null hypothesis at a 5% significance level? Circle one: **Yes No**
      Explain.
      Did your conclusion here match the one you made in part 4?

ILP: Is the Linear Model Appropriate? Are the Assumptions Met for Inference?

You will produce and examine the residuals from the regression line as well as create some plots to assess the fit of the linear model. This will also serve to evaluate the validity of the testing and confidence intervals performed in the earlier activities.

Regression assumptions may be stated in terms of the response variable or in terms of the error terms. The statistical model for simple linear regression assumes that for each value of x, the observed values of the response are normally distributed with some mean (that may depend on x in a linear way) and a standard deviation σ that does not depend on x. **For each x, *Y is* N(E(*Y*), σ), where E(*Y*) = $\beta_0 + \beta_1$x.**

Thinking about the error terms, we can say the **true error terms** (those that we do not observe) are the difference of the response and the true mean (for a given x). These errors are to have a normal distribution with a mean of 0 and a standard deviation of σ (that does not depend on x).

***Task*:** Obtain the residuals for the regression of Books on TV, and generate the appropriate plots to check the assumptions of the linear model.

1. Obtain the residuals in SPSS using **Analyze> Regression> Linear**, but instead of running it right away, click on the **Save** button. In the box that will open, select **Unstandardized** under the **Residuals** heading, and click **Continue**. This saves the residuals as a new variable. Now, construct a scatterplot of the residuals against the explanatory variable, TV. Sketch the general pattern of the plot.

   What assumption of the error terms does this plot help assess?

   What conclusion can you draw from this plot?

2. Construct a Q-Q plot of the residuals.
   Sketch the general pattern of the plot.

   What assumption of the error terms does this plot help assess?

   Based on the plot what is your conclusion about this assumption?

Cool-Down: Think About It

1. Why would you not want to use this model to predict the yearly number of books read for a patron that watches 15 minutes of television weekly?

2. The library plans to report a few prediction intervals to predict the number of books read in a year for patrons who watch a given amount of television per week. What amount of television watched weekly would produce a prediction interval with the narrowest margin of error?

3. Which is the only test statistic you can use if you want to test if the explanatory variable television is a significant *negative* linear predictor of the number of books read per week?

Example Exam Question on Regression

A doctor wanted to study the relationship between a male's age and HDL (so-called good) cholesterol. She randomly selects 18 male adult patients and records their x = age and y = HDL. A scatterplot of the data indicated an approximately linear relationship overall, so she performs a linear regression analysis using SPSS.

Model Summary

Model	R	R Square	Adjusted R Square	Std. Error of the Estimate
1	.598	.358	.318	6.807

ANOVA

Model		Sum of Squares	df	Mean Square	F	Sig.
1	Regression	413.01	1	413.01	8.91	.009
	Residual	741.43	16	46.34		
	Total	1154.44	17			

Coefficients

Model		Unstandardized Coefficients		Standardized Coefficients		
		B	Std. Error	Beta	t	Sig.
1	(Constant)	61.05	6.08		10.05	.000
	Age	-.38	.13	-.598	-2.99	.009

We also have: $\bar{x} = 46.22$ and $S_{xx} = \sum (x - \bar{x})^2 = 2883.131$.

a. What is the correlation between age and HDL cholesterol levels?

final answer: ______________________________

b. Provide the least square regression line for predicting HDL from age:

final answer: ____

c. Based on this model, predict HDL cholesterol for a male who is 30 years old.

final answer: ______________________________

d. Use a 1% significance level to assess if there is a significant linear relationship between the age and HDL cholesterol for male adults. State the hypotheses to be tested, the observed value of the test statistic, the corresponding *p*-value, and your decision.

Hypotheses: H_0:__________________ H_a:__________________

Test Statistic Value: __________________ *p*-value: ______________________

Decision: (circle) **Fail to reject H_0** **Reject H_0**

continued ...

Exam Question Example on Regression continued ...

e. Calculate a 95% confidence interval for the mean HDL cholesterol level for all 30-year-old male adults.

final answer: ______________________________

f. Consider the residual plot provided below. Does this plot support the conclusion that the linear regression model is appropriate? ***Yes*** ***No***

Explain:

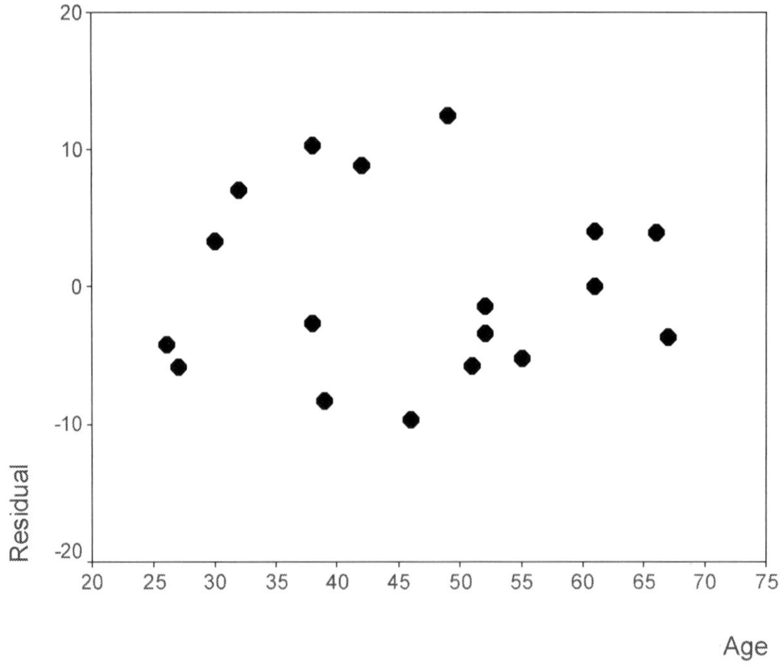

Lab 12: Chi-Square Tests

Objective: In this lab, you will learn how to perform three Chi-square tests (the test of goodness of fit, the test of independence, and the test of homogeneity) that are used to analyze categorical responses.

Overview: You will see three **Chi-Square tests**: the tests of **goodness of fit**, **independence**, and **homogeneity**. For all three tests, the data are generally presented in the form of a **contingency table** (a rectangular array of numbers in cells). All three tests are based on the **Chi-Square statistic:**

$\chi^2 = \sum \frac{(O_i - E_i)^2}{E_i}$, where O_i is the observed count and

E_i is the expected count under the corresponding null hypothesis.

Goodness of Fit Test: This test answers the question, "Do the data fit well compared to a specified distribution?" It considers one categorical response, and assesses whether the proportion of sampled observations falling into each category matches well to a specified distribution. The null hypothesis specifies this distribution which describes the population proportion of observations in each category.

Test of Homogeneity: This test answers the question, "Do two or more populations have the same distribution for one categorical variable?" It considers one categorical response, and assesses whether the model for this response is the same in two (or more) populations. The null hypothesis is that the distribution of the categorical variable is the same for the two (or more) populations.

Test of Independence: This test answers the question, "Are two factors (or variables) independent for a population under study?" It considers two categorical variables (sometimes one is a response and the other is explanatory), and assesses whether there appears to be a relationship between these two variables for a single population. The null hypothesis is that the two categorical variables are independent (not related) for the population of interest.

There are a few properties of the Chi-square distribution that you might find useful. The expected value of a Chi-square distribution is its degrees of freedom (mean = $\mu = df$), and its variance is 2 times its degrees of freedom. Thus, its standard deviation is the square root of 2 times the degrees of freedom ($\sigma^2 = 2 * df$ so $\sigma = \sqrt{2 * df}$). This frame of reference can help assess if our observed statistic is unusual under the null hypothesis or somewhat consistent with the null hypothesis.

Formula Card:

Chi-Square Tests	
Test of Independence & Test of Homogeneity	Test for Goodness of Fit
Expected Count $E = \text{expected} = \dfrac{\text{row total} \times \text{column total}}{\text{total } n}$	**Expected Count** $E_i = \text{expected} = np_{i0}$
Test Statistic $X^2 = \sum \dfrac{(O-E)^2}{E} = \sum \dfrac{(\text{observed} - \text{expected})^2}{\text{expected}}$ $\text{df} = (r-1)(c-1)$	**Test Statistic** $X^2 = \sum \dfrac{(O-E)^2}{E} = \sum \dfrac{(\text{observed} - \text{expected})^2}{\text{expected}}$ $\text{df} = k-1$
If Y follows a $\chi^2(df)$ distribution, then $E(Y) = df$ and $\text{Var}(Y) = 2(df)$.	

Warm-Up: Check Your Understanding

Fill in the blank with the name of the most appropriate Chi-square test to address the research question.

1. A researcher wants to determine if scoring high or low on an artistic ability test depends on being right or left-handed.

   Answer: Chi-Square test of ______________________________

2. A national organization wants to compare the distribution of level of highest education completed (high school, college, masters, doctoral) for Republicans versus Democrats.

   Answer: Chi-Square test of ______________________________

3. A preservation society has the percentages of five main types of fish in the river from 10 years ago. After noticing an imbalance recently, they add some fish from hatcheries to the river. How can they determine if they restored the ecosystem from a new sample of fish?

   Answer: Chi-Square test of ______________________________

ILP: Is There a Different Pattern in the Distribution of Accidental Deaths in a Certain Region Compared to the Pattern in the Entire United States?

Background: According to the records of the National Safety Council, accidental deaths in the United States during 2002 had the following distribution according to the principal types of accidents.

Motor Vehicle	Falls	Drowning	Fire	Poison	Other
45%	15%	4%	3%	16%	17%

Suppose that an accidental death data set from a particular geographical region yielded the following frequency distribution for the principal types of accidents:

Motor Vehicle	Falls	Drowning	Fire	Poison	Other
442	161	42	33	162	150

Do these data show a significantly different pattern in the distribution of accidental deaths in the particular region compared to the pattern in the entire United States? Use a 5% significance level. (Source: National Safety Council Website, 2005)

Task: Perform a Chi-square goodness of fit test to asses whether the data fit well with the model specified in the null hypothesis.

1. **State the null hypothesis:**

   H_0: ____

   Determine Alpha: We were told to use a significance level of 5%.

> **Remember:** Your hypotheses and parameter definition should always be a statement about the **population(s)** under study.

2. **Checking the Assumptions**
   a. Find the expected counts and fill them in the table below.

	MotorVehicle	Falls	Drowning	Fire	Poison	Other	Total
Null %	45%	15%	4%	3%	16%	17%	100%
Observed	442	161	42	33	162	150	990
Expected							

   b. Do all cells have expected counts greater than 5? **Yes** **No**

3. **Compute the Test Statistic and Calculate the *p*-value**

Test-Statistic

a. Complete the calculation of the test statistic based on your table above.

$$X^2 = \frac{(442-445.5)^2}{445.5} + \frac{(161-148.5)^2}{148.5} + \frac{(42-39.6)^2}{39.6}$$
$$+ \frac{(33-29.7)^2}{29.7} + \frac{(162-158.4)^2}{158.4} + \frac{(\ldots\ldots\ldots\ldots\ldots)^2}{} = 3.663$$

Calculate the *p*-Value:

b. Based on the output, answer the following:

Test Statistic

	Category
Chi-Square	3.663
df	5
Asymp. Sig.	.599

PD-INEL

i. The *p*-value is ______________ .

ii. The expected value of the test statistic assuming H0 is true is ____________

iii. The large *p*-value is consistent with the fact that our observed test statistic value is **greater than** **less than** the expected test statistic value (under the null hypothesis).

4. **Evaluate the *p*-value and Conclusion:**

Evaluate the *p*-value:

What is your decision at a 5% significance level? **Reject H_0** **Fail to Reject H_0**

> **Remember:** Reject H_0 ⇔ Results statistically significant
> Fail to Reject H_0 ⇔ Results not statistically significant

Conclusion:

What is your conclusion *in the context of the problem*?

ILP: Comparison of the Distribution of Academic Degrees: Males Versus Females

Background: How do women and men compare in the pursuit of academic degrees? The table presents counts (in thousands) from the Statistical Abstract of degrees earned in 1996 categorized by the level of the degree and the sex of the recipient.

	Bachelor	Master	Professional	Doctorate
Female	642	227	32	18
Male	522	179	45	27

Task: Perform a Chi-square test of homogeneity. Use a 1% significance level.

Hypothesis Test:

1. State the null hypothesis:

H_0: ____

____

Determine Alpha: We were told to use a significance level of 1%.

2. Checking the Assumptions

Sex * Degree Crosstabulation

			Degree				
			Bachelor	Master	Professional	Doctorate	Total
Sex	Female	Count	642	227	32	18	919
		Expected Count	632.2	220.5	41.8	24.4	919.0
	Male	Count	522	179	45	27	773
		Expected Count	531.8	185.5	35.2	20.6	773.0
	Total	Count	1164	406	77	45	1692
		Expected Count	1164.0	406.0	77.0	45.0	1692.0

a. Show how the expected count 531.8 (first cell for males) was computed.

b. Based on the table, do the assumptions appear to be met to perform the test? (Are all expected counts greater than 5?) **Yes** **No**

3. **Compute the Test-Statistic and Calculate the *p*-value**
   Test-Statistic
   a. What is the value of the test statistic? ____ = ____________

Chi-Square Tests

	Value	df	Asymp. Sig. (2-sided)
Pearson Chi-Square	9.514[a]	3	.023
Likelihood Ratio	9.485	3	.023
Linear-by-Linear Association	5.099	1	.024
N of Valid Cases	1692		

a. 0 cells (.0%) have expected count less than 5. The minimum expected count is 20.56.

   Calculate the *p*-value:
   b. Based on the output, the *p*-value is ____________ .

4. **Evaluate the *p*-value and Conclusion**
   Evaluate the *p*-value:
   What is your decision at a 1% significance level? **Reject H_0** **Fail to Reject H_0**

   Conclusion:
   What is your conclusion at a 1% significance level *in context of the problem*?

5. Would your decision and conclusion change if the significance level was:

   - 5% instead of 1%?
   - 3% instead of 1%?
   - 2.3% instead of 1%?
   - 2% instead of 1%?

   Based on your answers, the *p*-value represents the ________________ significance level at which the results would be statistically significant.

Cool-Down: Name That Scenario

Match each research question with the appropriate Chi-Square test that should be used to answer the question

1. Is student status (in-state versus out-of-state) associated with one's eventual graduation outcome (graduating versus not graduating)?

   Answer: Chi-Square test of ______________________________

2. To test a theory that people have no preference among four different outdoor activities, you ask 100 people to select among jogging, bicycling, hiking, or swimming.

   Answer: Chi-Square test of ______________________________

3. A biostatistician would like to determine if the ratio of the blood type in the storage for transfusions should be different in Hawaii from the main land. She collected a sample of blood types of 10,000 people in Hawaii and that of 100,000 people in the mainland. She wishes to see if the breakdown of blood types (A, B, AB and 0) is the same for both populations.

   Answer: Chi-Square test of ______________________________

Example Exam Question on Chi-Square Tests

A study was performed to examine the attendance pattern and exam performance of students in an intro statistics course. A sample of 96 students enrolled in an intro stat course was selected (these can be considered a random sample of such students). These 96 students were classified by their attendance status (regularly attend lecture or not) and their performance on the midterm exam, classified as low (below 50%), middle (50% to 80%), and high (above 80%).

Regularly Attend? * Exam Performance Crosstabulation

Count

		Exam Performance			Total
		Low	Middle	High	
Regularly Attend?	Yes	12	20	28	60
	No	16	12	8	36
Total		28	32	36	96

Chi-Square Tests

	Value	df	Asymp. Sig.
Pearson Chi-Square	8.195 [a]	2	.017
Likelihood Ratio	8.298	2	.016
Linear-by-Linear Association	8.067	1	.005
N of Valid Cases	96		

a. 0 cells (.0%) have expected count less than 5. The minimum expected count is 10.50.

a. Give the name of the chi-square test for assessing if there is a relationship between attendance status and exam performance.

b. Based on the above data, what proportion of regular attendees performed high on the exam?

Final answer: ____________________

c. Assuming there is no relationship between attendance status and exam performance, how many regular attendees would you expect to perform high on the exam? ***Show all work.***

Final answer: ____________________

d. Assuming there is no relationship between attendance status and exam performance, what is the distribution of the test statistic?

Final answer: ____________________

e. Use a level of 0.05 to assess if there is a significant relationship between attendance status and exam performance.

Test Statistic Value: ________________ *p*-value: ________________

Thus, there (circle your answer): **does** **does not** appear to be an association between attendance status and exam performance.

Exam 1 Questions

Statistics 250 Fall 2011 Exam 1

1. **A medical study** will be conducted and the following data will be recorded for the subjects of the study. **[7 points]**

a. Classify each of the following types of data as quantitative or categorical (circle your answer).

Smoker (yes or no)	***Quantitative***	***Categorical***
Body temperature (degrees Fahrenheit)	***Quantitative***	***Categorical***

b. The systolic blood pressures (mm of mercury) for the subjects in the Study (referred to as Study 1) are to be compared to the blood pressure results for the subjects of a second similar medical study (Study 2). The side-by-side boxplots for blood pressure are provided.

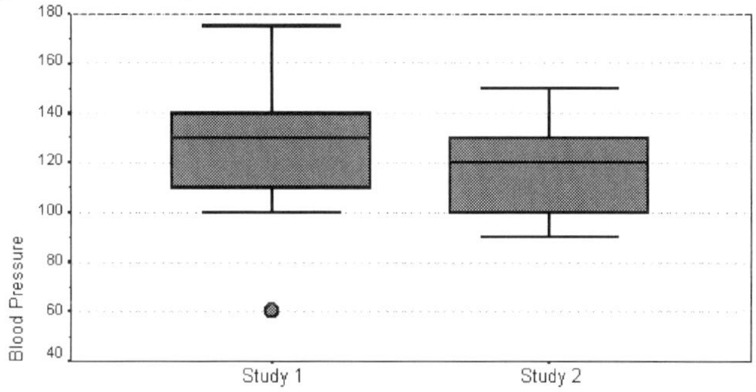

Clearly fill in the blank to complete the following statements.

i. The person with the lowest blood pressure in Study 1 has a blood pressure of __________ mm of mercury.

ii. About 75% of the participants in Study 2 have a blood pressure higher than __________ mm of mercury.

iii. Select the name of numerical quantity represented by the boxplot that completes this sentence:

The *median* for blood pressures in Study 1 is (roughly) equal to

______________________ for the blood pressures in Study 2.

c. A Q-Q plot of the Study 2 blood pressures is given at the right. Does a normal distribution appear to be a reasonable model for blood pressures for the underlying population?

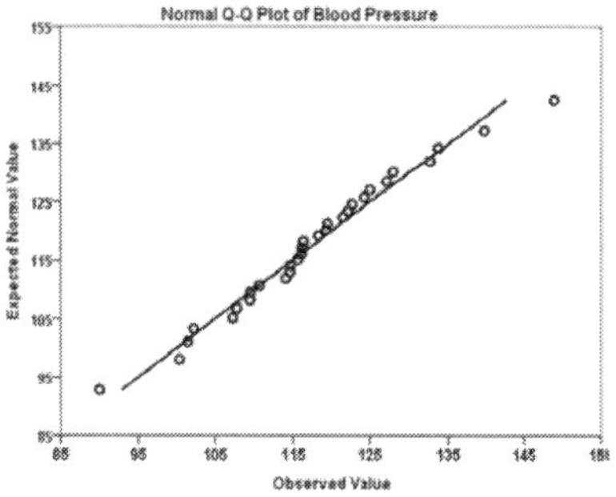

Circle one: ***Yes*** ***No***

Briefly explain:

2. **Another medical study** (Study 3) found that people who drink more than 4 cups of coffee a day have more heart attacks than people who drink less coffee or no coffee. This lead some doctors to suspect that coffee may be a contributing factor in causing heart attacks. However, more careful analysis of the data showed that heavy coffee drinkers tend to smoke more than other people. This is an example of: (circle one)
   [2 points]
   i. the placebo effect.
   ii. confounding.
   iii. response bias.
   iv. the double-blind technique.
   v. a completely randomized design.

---

3. **Giving Doctors:** Researchers chose 60 communities at random, and then chose some doctors at random in each community. In total, they interviewed 10,881 doctors. Overall, 77.3% of the doctors in the study said they had given some care free or at reduced rates because of the patient's financial need in the month before the interview. **[4 points]**

a. The number 77.3% is: (clearly circle your answer)
   i. a statistic, because it is a numerical summary for a sample.
   ii. a statistic, because it is a numerical summary for a population.
   iii. a parameter, because it is a numerical summary for a sample.
   iv. a parameter, because it is a numerical summary for a population.

b. Some doctors who did not give any charity care may say that they did. If so, the study suffers from:
   (clearly circle your answer)
   i. a large margin of error.
   ii. non-response bias.
   iii. response bias, the sample result will systematically overestimate the true percent of doctors who give charity care.
   iv. response bias, the sample result will systematically underestimate the true percent of doctors who give charity care.

---

4. **Laptops:** If you wanted to estimate the proportion of Ann Arbor High School students that bring a laptop to school, at **99%** confidence level with a margin of error of (at most) 4%, how many students should be selected? Show your work. **[3 points]**

Final answer: ______________________

5. **Food Plan Comparisons:** Various food plans are available for residents of a large community starting at $10 a day up to $30 a day, in increments of $5. Below provides a summary of the distribution of daily food plan cost for 100 residents that live in Building A of this community and for 100 residents that live in Building B of this community.

Daily Cost	**$5**	**$10**	**$15**	**$20**	**$25**
Number of Building A residents	30	15	10	15	30
Number of Building B residents	20	0	60	0	20

So 30 of the 100 residents from Building A had the $5 a day meal plan, and 20 of the 100 residents from Building B had the $25 a day meal plan. Note: Questions can be answered without computations. **[4 points]**

a. Which set of results has the larger average daily cost? (circle your answer)

Building A ***Building B*** ***The averages are equal*** ***Cannot be determined***

b. The standard deviation for the Building A results __________ the standard deviation for the Building B results. (clearly circle your answer)

is greater than ***is equal to*** ***is less than*** ***cannot be compared to***

---

6. **Cold Season:** The close quarters in dormitories, apartments, and cafeterias help to spread the cold virus easily, especially in fall and winter months. According to health professionals, it's not unusual for college students to have 2 to 3 colds in a normal academic year. The table below displays the (partial) distribution for the variable X = the number of colds college students have in a given academic year. **[7 points]**

X = # colds	**0**	**1**	**2**	**3**	**4**
Probability P(X=x)		**.15**	**.25**	**.30**	

a. Suppose a randomly selected student is twice as likely to get 4 colds as he/she is to get no (0) colds. Using this information, complete the above distribution.

b. On average, how many colds can a college student expect to get in a given academic year? Write the correct symbol, provide the value and units. Show your work.

__________ = ____________________

symbol

c. Jane is a college student who has already had one cold in September and another in October. What is the probability she will get just one more cold before the end of the academic year (next April)? Show your work.

Final answer: ____________

7. **True or False? [4 points]** For each statement, clearly circle your answer.

a. A study reported a 95% confidence interval for a population proportion of all Americans who attend religious services regularly to be (0.43, 0.66). If the study were repeated a large number of times, then the interval (0.43, 0.66) will capture the population proportion in approximately 95% of the studies. ***True*** ***False***

b. Statistics can vary from sample to sample, whereas parameters are fixed values. ***True*** ***False***

c. We should decide on the significance level α before looking at the data.
True ***False***

d. A QQ plot determines whether a sample is normal.
True ***False***

---

8. **Blackberries:** Suppose that 40% of all UM faculty members own a Blackberry cell phone. A random sample of 300 UM faculty members will be selected and the sample proportion $\hat{p}$ that own a Blackberry cell phone will be determined. **[5 points]**

a. What are the possible values for the sample proportion $\hat{p}$? Make a detailed sketch of the distribution of all such $\hat{p}$ values if repeated random samples of 300 UM faculty members were obtained. Provide the name of the distribution with necessary values to completely specify the distribution.

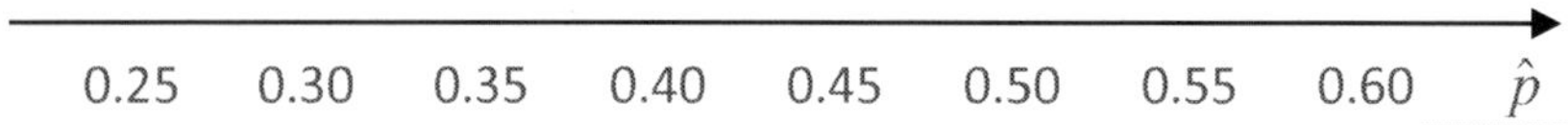

b. What is the probability of seeing a sample proportion as high as 0.45 or even higher? Show your work.

Final answer: ________________

9. **Barefoot Running Shoes:** A study to examine the injury potential of barefoot running shoes focused on how female runners adapted to barefoot running shoes. The study found that 50% failed to adapt their running style to accommodate the different choice of footwear. Assume this 50% rate is correct for the population of all female runners. Suppose a small follow-up study will be conducted by the company that makes the shoes. A random sample of 12 female runners will be selected and X will represent the number of female runners in the sample who do adjust their running form. **[6 points]**

a. What is the distribution of X? Provide all details.

Distribution: ______________________________

b. For the follow-up study, what is the probability that at least 11 of the 12 randomly selected female runners will adjust their running form? Show all work.

Final answer: ______________

c. For the follow-up study, how many of the 12 randomly selected female runners are expected to adjust their running form? Include the units.

Final answer: ____________________

---

10. **Healthier Snacks:** Recently, many schools have started to offer healthier options in vending machines to their students. Rather than selling chips, soda, and candy, new vending machines are selling things like baby carrots, hummus, and hard boiled eggs. While some students embrace the changes, others aren't so sure, even those who come from families who encourage healthy snacks at home. At a certain school ...

- 65% of students have families who encourage healthy snacks at home **(A = "At Home")**
- 25% of students regularly choose snacks from the healthy vending machines
- 15% of students have families who encourage healthy snacks at home *and* regularly choose snacks from the healthy vending machines.

Are the events "having a family that encourages healthy snacks at home" and "regularly choosing snacks from the healthy vending machine" independent events? **[3 points]**

Circle one: ***Yes*** ***No***

Provide numerical support for your answer.

11. Pumpkins: A farmer who sponsors a Jack-O-Lantern carving contest supplies a pumpkin for each of the contest participants. The participants are grouped based on experience level (novice, experienced, expert) and size of pumpkin that they received (small, medium, large, extra-large). The pumpkins that are provided weigh an average of 15 pounds with a standard deviation of 4 pounds and can be assumed to follow approximately a normal distribution. **[10 points]**

a. Joe is participating in the Jack-O-Lantern carving contest. He would like to carve a 20 pound pumpkin. Based on the model, what is the probability he receives a pumpkin weighing exactly 20 pounds?

Final answer: ______________________

b. The standard deviation for weights of the supplied pumpkins is stated as 4 pounds. Provide a concise sentence that appropriately interprets this value of 4 as an average distance.

c. If the "extra-large" pumpkins are the 15% of pumpkins that weight the most pumpkins, how much must a pumpkin weigh to fall into the "extra-large" category? Report to the nearest tenth of a pound.

Final answer: ______________________

d. The time it takes a participant to carve his or her pumpkin is uniformly distributed between 5 and 45 minutes.

   i. Provide a complete sketch of this distribution. Include values and labels on each axis.

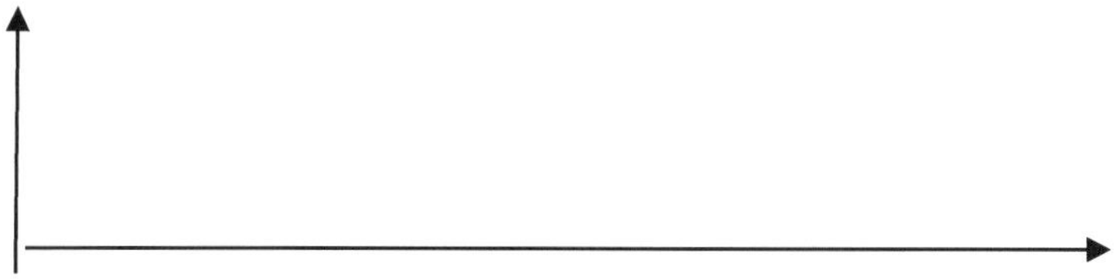

   ii. What is the probability a randomly selected participant will take more than half an hour to carve their pumpkin? Show your work.

Final answer: ____________________

---

12. Nausea: A certain pain-relief medication causes nausea in 25% of the patients taking it. A new formulation of this medication has been developed that should *reduce* this proportion. The research team would like to test this theory about *p*, the proportion for the population of all patients taking the new formulation who become nauseated. There are 50 randomly selected patients that will receive the new formulation as part of this study. **[12 points]**

a. This study would be considered an (circle one) ***experiment*** ***observational study***.

b. State the hypotheses to be tested in terms of *p*.

H_0:____________________ H_a:____________________

c. The level of significance for the test is set at 5%. Is the following statement **True or False**?

For this study, a consequence of a Type I error will be that a new formulation that does not decrease nausea may be falsely advertised and sold to patients.

Circle one: ***True*** ***False***

d. One data condition is that the sample size is large enough to use an approximate normal model for the sample proportion. Provide the numerical justification to verify that this condition is satisfied.

e. Study results: Of the 50 patients, a total of 11 reported nausea after taking the new formulation. Compute the corresponding test statistic value.

Test statistic value: ______________

f. Provide a complete sketch of the corresponding *p*-value (including a *sufficiently* labeled curve and x-axis), and report the value of the *p*-value.

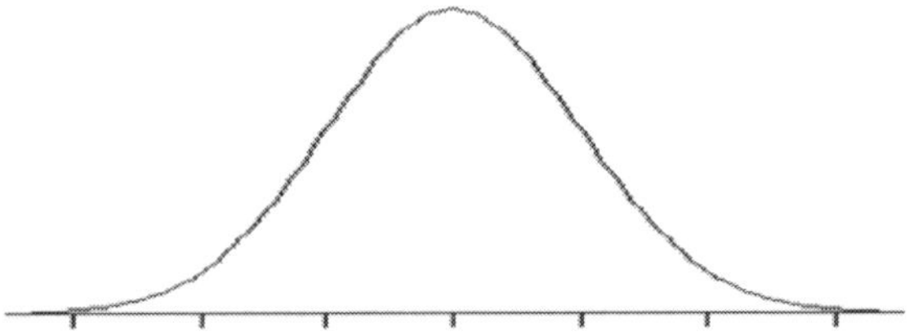

p-value: ______________

g. The appropriate decision is: (circle one) ***reject H_0*** ***fail to reject H_0***

h. The (real-world) conclusion would be: (circle one)

- ***The new formulation significantly lowers the incidence of nausea below 25% for the population of all patients represented by the sample.***
- ***There is insufficient evidence that the new formulation lowers the incidence of nausea below 25% for the population of all patients represented by the sample.***

13. iPads: Mark (a UM graduate student) selected a random sample of 200 UM faculty members and found that 44 (or 22%) own an iPad. To provide an indication of the accuracy of this point estimate, Mark would like to construct a corresponding 90% confidence interval estimate. **[8 points]**

a. To create the confidence interval, a standard error is needed. Compute that standard error.

Final answer: ____________________

b. **True or False**?

This standard error estimates the standard deviation for all the possible $\hat{p}$ values that would be obtained if repeated samples of 200 UM faculty members were taken.

Circle one: ***True False***

c. Provide the 90% confidence interval to estimate the population proportion of all UM faculty members that own an iPad. Show your work.

90% Confidence Interval: ______________ to ______________

d. If Mark had constructed a 95% confidence interval instead, that interval would have been (circle one) ***narrower than*** ***wider than*** ***the same width as*** the 99% confidence interval estimate.

e. Which of the following are appropriate interpretations of the 90% confidence interval in part c? (clearly circle your answer)

- We estimate with 90% confidence the population proportion of all UM faculty owning iPads to be in that confidence interval.
- There's a 90% chance that the population proportion of all UM faculty owning iPads is in that confidence interval.
- 90% of all possible confidence intervals are expected to contain the sample proportion $\hat{p}$.
- None of the above
- All of the above

Statistics 250 Winter 2012 Exam 1

1. **Test Prep Company:** Most students are required to take some sort of standardized test as part of the college application process. A certain test prep company claims that students who take their course do better on standardized tests than students who don't.

   The company provides the following graph of the test scores earned by students who did not take the test prep course (Group 1 = No) and test scores of scores of students who did take the course (Group 2 = Yes).

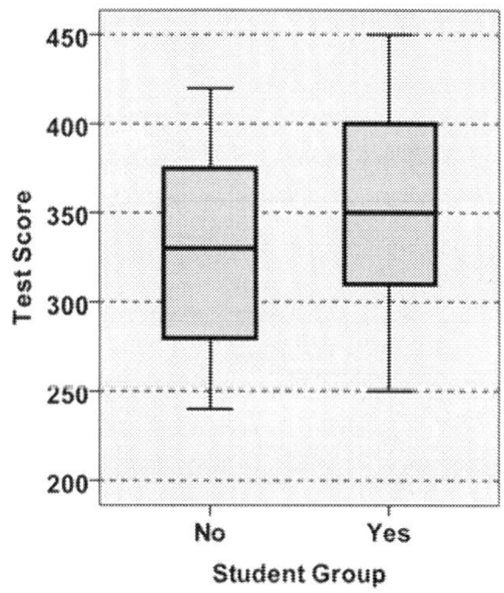

a. Complete the following sentences:
[2]

   i. Based on the median performance, students who ***do*** take their course score approximately ______________ points higher than students who **don't** take their course.

   ii. Of those that ***do*** take their course, 25% scored higher than _____________ points on the test.

b. A statistician is asked to compare these two sets of observations and perform a test of hypotheses. One assumption required for this inference to be valid is that the observations in each group need to be considered a random sample from a population that has a **normal** model for the response (scores).
[3]

   i. What graph is ***best*** for assessing if a normal model is valid for the population of test scores for students that do take their prep course?

   Final answer: ___________________________________

   ii. Sketch at the right what the graph of the test score data for the students that did take the prep course ***should*** look like if the assumption of normality is valid.

2. How does the amount of "quality time" spent with family influence the risky behavior in teens? A study at the University of Florida suggested that spending more quality time with their families results in U.S. teens having less involvement in risky behavior, such as binge drinking and using illegal drugs.

a. In this study the explanatory variable is: ______________________________

[2]

b. In order to gather data, researchers took a random sample of U.S. families with teenagers and recorded the amount of quality time that was spent together in a week and recorded the teens' involvement in various risky behaviors. Which of the following two words best describes this study? Clearly circle your answer.

[1] ***experimental*** ***observational***

c. Let X = amount of quality time per week (in hours) that a U.S. teenager spends with his/her family. The probability density curve is shown in the graph to the right. The median of this distribution is 9 hours. Which of the following three values is the ***mean*** of this distribution? Clearly circle your answer:

[2] ***10 hours*** ***9 hours*** ***7.3 hours***

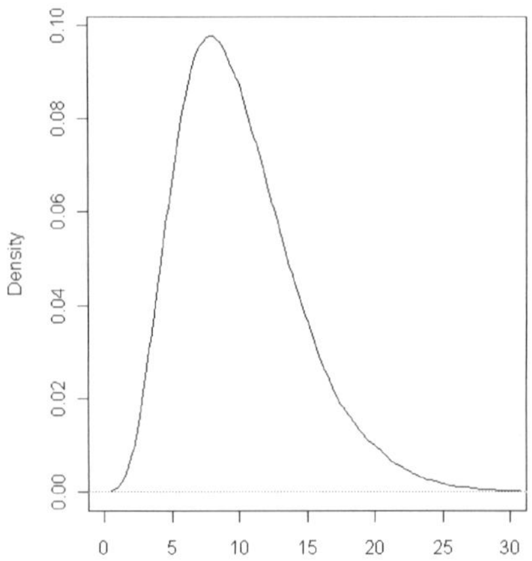

X = quality time a teen spends with family per week (in hours)

d. The standard deviation for X is about 4.5 hours. Which of the following is/are correct interpretation(s) of the standard deviation? Circle all that apply.

[3]

- The times vary from the mean by about 4.5 hours, on average.
- 68% of the times will fall within 4.5 hours from the mean amount of time.
- The average distance between the times and the mean is about 4.5 hours.

3. Satisfied Customers: A department store mails a customer satisfaction survey to people who make credit card purchases at the store. This month, 45,000 people made credit card purchases. Surveys are mailed to 1000 of these people, chosen at random. A total of 850 people returned the survey form of which 60% stated they were 'very satisfied' with their latest credit card purchase. **[4 points]**

a. In this survey, what is the population of interest? **Clearly circle your answer.**

- All shoppers at the department store.
- All people who make credit card purchases at the department store.
- All 1000 people to whom the survey was mailed.
- All 150 people who returned the survey form.

b. Which of the following terms applies to the value of 60%?
Clearly circle all that apply.

parameter ***population proportion*** ***statistic***

p $\hat{p}$ ***sample proportion***

---

4. Math Ability: Alysha scored 670 points on the Mathematics part of the SAT. The distribution of SAT Math scores in that year was normal with mean of 516 points and standard deviation 116 points. John took the ACT and scored 26 on the Mathematics portion. ACT Math scores for that year were normally distributed with mean of 21.0 points and standard deviation 5.3 points.

a. What percentile does Alysha's score correspond to? Show your work.
[2]

Alysha's score corresponds to the ______________________ percentile for the SAT Math distribution.

b. Assuming both tests measure the same type of ability, who had the higher score? Include numerical support.

[2] Circle one: ***Alysha*** ***John*** had the higher score because:

c. All students who took the SAT Math test and scored in the top 2% received a letter from a national honor society inviting them to join. What is the lowest SAT Math score needed to be invited? Show your work.
[2]

You need a score of at least ________________ points to be invited.

5. **Being Productive:** The National Center for Productivity released information about the efficiency of workers. The graph at the right showing the productivity index measured over the past 12 months was included in the report. When presented with this graph, one manager stated ***"So, the distribution of productivity is skewed to the left."*** Is this an appropriate statement based on this graph?

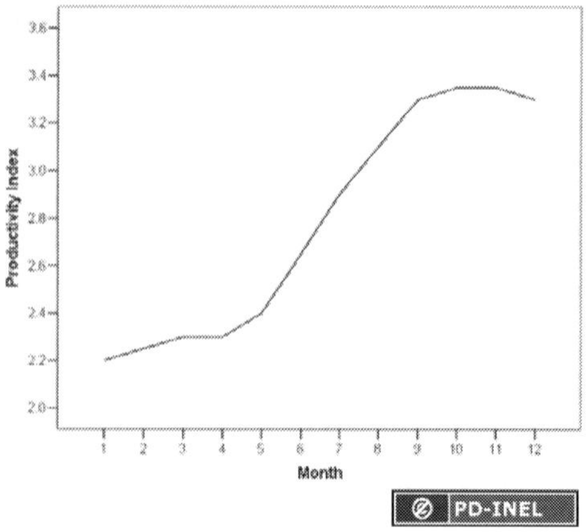

[2] Circle one: ***YES*** ***NO***

Briefly explain:

---

6. **Blood Types:** In the United Kingdom (UK), 35% of the population has Type A positive blood.

a. Consider a random sample of 9 people from the UK. What is the probability that exactly 2 of the 9 people have Type A positive blood?

[2]

Final Answer: ____________________

b. Now consider a random sample of 90 people from the UK. How many would you *expect* to have Type A positive blood?

[2]

Final Answer: ____________________

c. Again consider taking a random sample of 90 people from the UK. Suppose exactly 40 do have Type A positive blood. How many standard deviations is this observed result of 40 from what is expected?

[2]

Final Answer: ____________________

d. You plan to *estimate* the probability of seeing at least 40 with Type A positive blood in the random sample of 90 people from the UK. **Verify** that the sample size is indeed large enough to use *normal approximation*.

[1]

7. Relationship between Eye Color and a Certain Ophthalmic Trait: Three hundred randomly selected individuals are studied as to either eye color (blue, brown, or other) is related to whether or not the individual has a particular ophthalmic trait (yes or no). The following results were recorded and are assumed to accurately reflect the corresponding population.

a. What is the probability that blue eyed people have this particular trait? (Report answer to two decimal places.)

[2]

	Eye Color			
Trait?	**Blue**	**Brown**	**Other**	***Total***
Yes	70	30	20	***120***
No	20	110	50	***180***
Total	***90***	***140***	***70***	***300***

Final Answer: ____________________

b. To assess whether the event of being **blue-eyed** was *independent* of the event of **having this particular trait**, you would compare the probability requested in part (a) with another probability. What is that *value*?

[1]

Final Answer: ____________________

8. Florida Hurricanes: The following distribution for X = *annual number of hurricanes in the state of Florida* was reported in an article entitled "What Seasonal Hurricane Forecasts Mean to Florida".

X = # Hurricanes	**0**	**1**	**2**	**3**	**4**
Annual Probability	**0.60**	**0.30**	**0.07**	**0.02**	**0.01**

a. Based on this model, what is the probability of there being *at least* one hurricane in the state of Florida over the next year?

[1]

Final Answer: ____________________

b. Given that two hurricanes have already occurred and the year is not yet over, what is the probability that there will actually be a total of three hurricanes in all?

[2]

Final Answer: ____________________

c. Based on this model, what is the annual *expected* number of hurricanes in the state of Florida? Provide the symbol and the value.

[3]

Final Answer: __________=__________________

9. Employed Graduates: According to the administration at a particular college, 55% of graduates are *typically* employed full time after graduation. To assess the impact of the current job market, the administration will take a survey of 619 *recent* graduates.

a. If the population proportion of all *recent* graduates employed full time is the same as the stated proportion *typically* employed, what is the distribution of the possible sample proportions the administration may obtain from its survey? Provide a complete sketch including a label and values on the x axis and a full label for the distribution.

[4]

b. The administration finds that 52% of the surveyed *recent* graduates were employed full time. If the population proportion of all *recent* graduates employed is the same as the stated proportion *typically* employed, how likely is it that the administration would have obtained a sample proportion as low as or lower than it obtained from its survey? Show all work.

[2]

Final Answer: ____________________

---

10. Sharing Passwords: Teenagers sharing online passwords with friends or significant others has become relatively common, according to a report in the NY Times. From a recent survey of 770 teenagers (aged 12 to 17 years old), 30% reported having shared a password with a friend or significant other.

a. Compute the corresponding standard error of the estimated proportion of all teenagers who have shared a password with a friend or significant other. Show your work.

[2]

Final Answer: ____________________

b. Provide a 99% confidence interval to estimate the population proportion of all teenagers who have shared an online password with a friend or significant other. (Report the interval out to two decimal places.)

[2]

Final Answer: (____________ , ____________)

c. The following statements are being considered for inclusion in the report to summarizing the study results. Clearly circle all statements that are correct and do apply.

[6]

- If this procedure were repeated many times, 99% of the resulting confidence intervals would contain the true population proportion of all teenagers who have shared an online password.
- If this procedure were repeated many times, we would estimate that the resulting sample proportions differ from each other by about the value in part (a).
- With 99% confidence, we would estimate the population proportion of all teenagers who have shared an online password falls in the interval computed in part (b).
- If this procedure were repeated many times, 99% of the resulting confidence intervals would contain the sample proportion of teenagers who have shared an online password.
- There is a 99% chance the true population proportion of all teenagers who have shared an online password falls in the interval computed in part (b).
- If a conservative 99% confidence interval had been made using the same survey results, the margin of error would have been smaller than that in part (b).

d. Another news organization requests a new study be done so that a 99% confidence interval can be reported with a margin of error of only 3%. How many people would a researcher need to sample to obtain such a confidence interval? Show all work.

[3]

Final Answer: ____________________

11. In-State or Out-of-State? In 2010, it was stated that 64% of the University of Michigan's incoming freshmen were from the state of Michigan (called 'in-state'). For a research project, an undergraduate student, Jessie, proposes to assess if the current proportion for the 2011 incoming freshmen class has changed from this previous level. She decides to use a 10% significance level.

a. State the appropriate hypotheses to be tested, including a clear definition of the parameter of interest.

[4] **H_0:**____________________ **H_a:**___________________________

where the parameter __________ **represents**_________________________

____

b. Jessie is still in the planning stages of her study. She comes to you (a student in Stats 250) to ask the following question: If the actual proportion of in-state UM freshmen in 2011 is not different from the previous 2010 level, what is the chance the test would conclude that the proportion has changed?

[1]

Final Answer: ____________________

c. Jessie has now taken a sample of 150 current UM freshmen and records their residency (in-state or out-of-state). She finds that 86 of the sampled freshmen are 'in-state'. Compute the appropriate test statistic value and find the *p*-value to test the hypotheses in part (a). Provide a complete sketch of the *p*-value; include a label and values on the x axis, a full label for the distribution, and the shaded area for the *p*-value.

[5]

Test statistic value: __________

***p*-value:** ________

d. The student should decide to (circle one) ***Reject H_0*** ***Fail to reject H_0***

[1]

e. Circle the correct word/set of words to complete the following conclusion:

[2] There (circle one) ***IS NOT*** ***IS*** sufficient evidence to say the (circle one) ***POPULATION PROPORTION*** ***SAMPLE PROPORTION*** of students from the state of Michigan in 2011 differs from that in 2010.

f. Based on Jessie's results, she could have made a Type II error.
Circle one: ***TRUE*** ***FALSE***

[1]

g. Later, it was discovered that Jessie only sampled UM freshmen from her own residence hall as it was easier for her to contact them. Complete the sentence:

[1] **This implies the study could be subject to** ________________________ **bias.**

Exam 2
Questions

Statistics 250 Fall 2011 Exam 2

1. **Parents help underage children lie to get on Facebook, survey finds** – The minimum age Facebook sets for its social network is 13. A recently released survey (November 1, 2011) shows parents are helping their children lie to get online. The survey, conducted by Harris Interactive, involved a random sampling of 645 parents age 26 and older with 10-12 year old children living with them. Of these 645 parents, 248 knew their underage children were signed up for Facebook.

a. Use the results of the survey to compute a *(general)* 99% confidence interval for the proportion of all such parents whose underage child is signed up for Facebook . (Show all work.)

[3]

Final answer: (_______________, _________________)

b. Which of the following is the correct interpretation of the 99% confidence *level*?

[2] Circle one:

i. The probability that the true proportion of all such parents with an underage child on Facebook falls between the bounds calculated in part (a) is 0.99.

ii. The probability that the sample proportion of parents with an underage child on Facebook falls between the bounds calculated in part (a) is 0.99 for repeated random samples of 645 parents.

iii. The true proportion of all such parents with an underage child on Facebook falls between the bounds calculated in part (a) 99% of the time.

iv. The true proportion of all such parents with an underage child on Facebook falls between the bounds of 99% of the possible confidence intervals resulting from repeated random samples of 645 parents.

c. How would the *conservative* 99% confidence interval compare to the interval calculated in part (a)?

[1] Clearly circle your answer: ***Narrower*** ***Same width*** ***Wider***

2. **True or False?** Determine if each of the following statements is true or false. Clearly circle your answer. **[1 point each]**

a. It would be appropriate to set the significance level α to 0.01 if incorrectly rejecting H_0 was not very consequential.

   True ***False***

b. If we "pool" variances from two samples for a common estimate of σ^2, then our pooled 95% confidence interval for $\mu_1 - \mu_2$ would be narrower than an unpooled 95% confidence interval.

   True ***False***

c. If a result is significant at the level $\alpha = 0.01$, then it would be significant at the level $\alpha = 0.05$.

   True ***False***

---

3. **Proposal C –** One proposal being voted on in an election is Proposal C: Should a tax be authorized for street and bridge resurfacing and reconstruction? Mr. Young is conducting a survey among residents of the city to see what proportion of residents will favor the proposal. Mr. Young has randomly selected 14 residents. Of the 14 residents, 13 are in favor to the proposal.

a. Mr. Young wishes to test the following hypotheses: H_0: $p = 0.70$ vs. H_a: $p > 0.70$, where p is the proportion of all residents in the city who favor the proposal. Perform the test based on Mr. Young's results using a 1% significance level. Report the value of the *p*-value and circle your decision. (Show all work.)

[4]

p-value: ____________________ Decision: ***Fail to reject*** H_0 ***Reject*** H_0

b. It has been recommended to repeat this survey using a larger sample size. For the new survey design (compared to the original study design), the probability of concluding more than 70% of all city residents favor the proposal *when there really is more than 70%* would ...

[2] Circle one: ***increase*** ***decrease*** ***stay the same***

The statistical name for this probability is: ____________________________

4. Statistically Significant – A researcher will compare UM students and MSU students on 50 different quantitative variables. He will use a 10% significance level to carry out the 50 independent samples t-tests (one for each variable). If, for each test, the null hypothesis is actually true, about how many "statistically significant" results will be produced?

[2]

Final answer: ______________________

---

5. Subaru's Management – Subaru has always been considered a "flat" company, that is, having fewer employees in management than other automobile manufacturers. Corporate headquarters recently reported that 14.4% of all of its employees were in management.

[1]

This 14.4% should be considered a:

(circle one) ***parameter*** ***statistic***.

6. **Psychology Study** – In a study of the relationship between temperament and personality, a random sample of 70 female high school students who had a high level of reactivity (HRL) and an independent random sample of 80 female high school students who had a low level of reactivity (LRL) were obtained. These students were then classified according to their attitude toward group pressure, recorded as resistant or submissive. Results are shown.

	Attitude		
Reactivity	**Resistant**	**Submissive**	**Total**
1 = HRL	21	49	70
2 = LRL	40	40	80

The researcher would like to assess if these results indicate a difference in the likelihood of a ***resistant*** attitude for the two reactivity populations, namely, to test: $H_0: p_1 = p_2$ versus $H_a: p_1 \neq p_2$

a. If the null hypothesis is true, there is just one common population proportion. Give an estimate of that common population proportion. Show all your work.

[2]

Final answer: ______________________________

b. The resulting test statistic value is -2.49. Provide the corresponding *p*-value. Show all work.

[2]

p-value: ______________________________

c. Which of the following is the appropriate statistical decision and conclusion at the 5% level? **Check one**.

[2]

____ Reject H_0 and conclude there is not a significant difference in the likelihood of a resistant attitude for the two reactivity populations.

____ Reject H_0 and conclude there is a significant difference in the likelihood of a resistant attitude for the two reactivity populations.

____ Fail to reject H_0 and conclude there is not a significant difference in the likelihood of a resistant attitude for the two reactivity populations.

____ Fail to reject H_0 and conclude there is a significant difference in the likelihood of a resistant attitude for the two reactivity populations.

d. What type of error could the researcher have made? Clearly circle one.

[1] ***Type I Error*** ***Type II Error*** ***Cannot be determined***

7. **How old is your Professor? –** Consider the population of all full professors at a very large university. The mean age for this population is 48 years and the standard deviation is 5 years. A random sample of 100 full professors will be taken and the sample mean age will be computed.

a. Give the mean and standard deviation for the distribution of the sample mean age based on a random sample of 100 full professors. Include the units.

[2]

Mean : ______________ Standard deviation: ____________________

b. What is the approximate probability that the sample mean age will differ from the true population mean age by more than 1 year? Show your work.

[2]

Final answer: ____________________

c. What is the name of the result that allows you to answer part (b), even though you were not told whether the population of ages followed a normal distribution?

[2] Result Name: ____

8. **How big is your Turkey? –** A 95% confidence interval for the average weight of a Thanksgiving turkey (in pounds) is given by (8.7, 21.3) based on a random sample of 50 households.

a. What is an estimate of the population average weight of a Thanksgiving turkey? Provide the symbol, the value, and units.

[3]

_______ = __________________________

b. In order to make the confidence interval of (8.7, 21.3) certain conditions are to be met. One condition is that the households sampled should be a random sample of all households, and that condition is satisfied. **Clearly** state the additional condition that is needed (be specific).

[2]

c. A Turkey company is planning their production levels assuming the average weight of Thanksgiving turkeys will be 18 pounds. Based on the 95% confidence interval is this population mean of 18 pounds a reasonable value?

[2] Decision: (circle one) ***Yes*** ***No***

Briefly explain your answer:

9. Step Forward – The Acculturation Rating Scale (ARS) measures the extent to which Mexican Americans have adapted to our culture with scores ranging from 0 to 50, higher scores imply better ability to adapt. During the summer of 2010, twenty Mexican American children (aged 10-14) who have been in the U.S. for only a few years participated in a Department of Education training program in Alabama. The program called "Step Forward" was designed to enhance the transition process to our culture. The ARS test was given to the children in early July just before they began the program and then given again in late October to assess whether the program had *increased* their ability to adapt to U.S. culture. The Department of Education would like to test this theory using a 5% level of significance. A partial listing of the data is provided along with some numerical summaries.

Subject	July ARS	Oct ARS	Diff = Oct - July
1	20	21	1
2	25	19	-6
3	19	23	4
. . .			
20	22	29	7
Mean	25.31	28.22	2.91
Standard Deviation	13.11	12.84	8.35

a. For this situation, specify the null and alternative hypotheses.

[2] H_0: ______________________ H_a: ______________________

b. The observed test statistic value is 1.56 and the corresponding *p*-value is 0.067.
[3]

i. What distribution was used to compute the *p*-value of 0.067? Be specific.

Final answer: ______________________

ii. Consider the following statement regarding this *p*-value.
If this study were repeated many times, we would see a t statistic as large or larger than 1.56 in about 6.7% of the repetitions.

Briefly state what is missing in order for this statement to be correct.

c. If the program really had no effect on average, what would have been the expected value of the test statistic?

[2]

Final answer: ______________________

10. What is that *p*-value? A market research firm was interested in assessing if a majority of all young adults like the package design for a new snack bar, that is, testing H_o: $p = 0.50$ versus H_a: $p > 0.50$ at a 5% level. A large random sample of young adults was shown the package design and the sample proportion that liked it was 0.48.

[2] Which of the following is/are reasonable for the *p*-value? Circle all that apply.

Circle one: ***0.65*** ***0.48*** ***0.05*** ***0.02***

---

11. Nickel Batteries – Cell phones of a particular brand have used nickel-cadmium batteries. A manufacturer has developed a new nickel-metal hydride battery. In a study to compare the two batteries (1 = cadmium, 2 = hydride) as to "talking time" (number of minutes before recharging), independent random samples of batteries for each battery type were tested in hopes that the average talking time for the new hydride battery was ***longer*** than for the old cadmium type. The following SPSS output was provided.

Group Statistics

	Battery	N	Mean	Std. Deviation	Std. Error Mean
Talk_Time	1=Cadmium	25	70.75	14.05	2.81
	2=Hydride	20	80.76	13.60	3.04

a. The hypotheses will be about μ_1 and μ_2. The null hypothesis is given. Provide the appropriate alternative hypothesis and a *complete* definition for μ_1.

[3] H_0: $\mu_1 = \mu_2$

H_a: ______________________________

where μ_1 = ______________________________

____

b. The research assistant who was generating the SPSS output created the three Q-Q plots shown below.

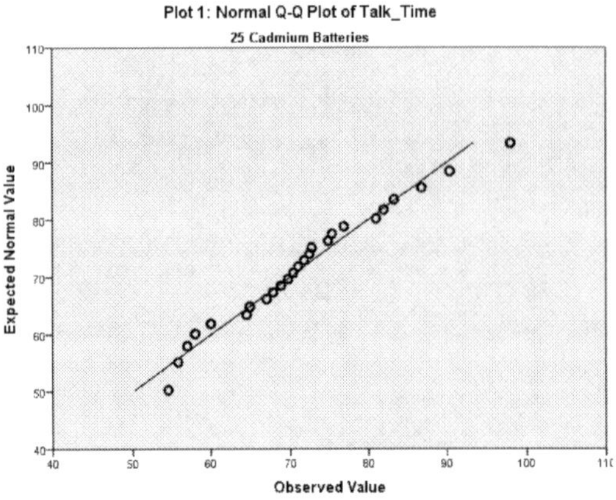

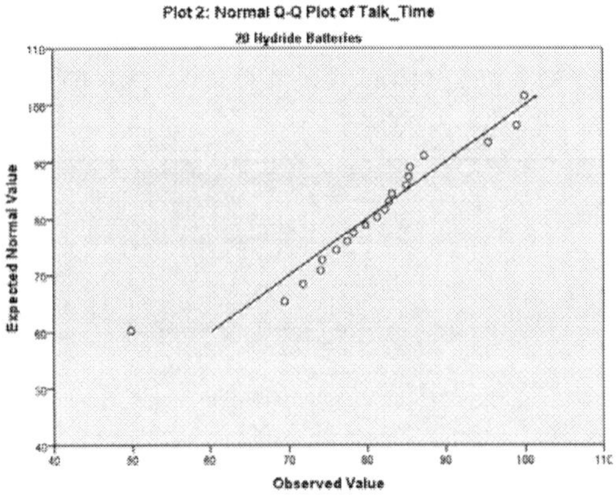

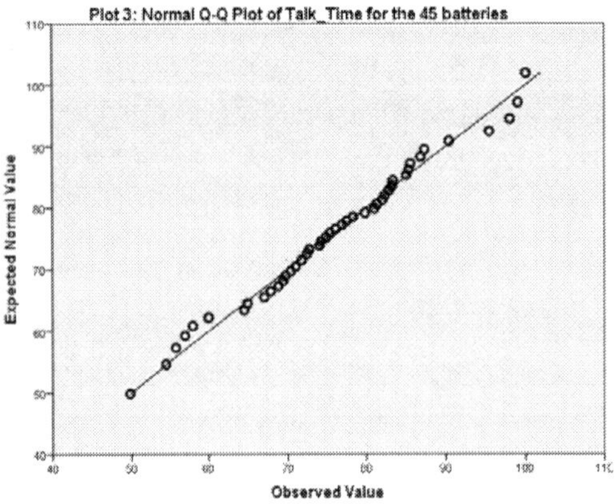

Which Q-Q plot or plots would be used for checking the assumptions for this test? [2]
Clearly circle your answer.

i. Plot 1 and Plot 2

ii. Plot 3

iii. All three Q-Q Plots

iv. None of them. The 45 talk-time measurements are large enough to not need to check normality.

Problem 11 Continued:

c. The following output proves a p-value = 0.700.

Independent Samples Test

		Levene's Test for Equality of Variances		t-test for Equality of Means						
									95% Confidence Interval of the Difference	
		F	Sig.	t	df	Sig. (2-tailed)	Mean Difference	Std. Error Difference	Lower	Upper
Talk_ Time	Equal variances assumed	.151	.700	-2.409		.0204	-10.01	4.16	-18.39	-1.63
	Equal variances not assumed			-2.418		.0201	-10.01	4.14	-18.37	-1.65

[2] Using appropriate symbols, write the null hypothesis that this p-value is used to test. H_0: ______________________

d. The researcher decided to use the pooled t-test results. Give the pooled *standard deviation* estimate. Show your work and express your answer out to two places after the decimal.

[2]

Final answer = ______________________

e. The df column is missing in the display in part (c).
What are the df corresponding to the "equal variances assumed" t-test?

[1]

df = ______________

f. For the hypotheses in part (a), provide the appropriate test statistic value and corresponding p-value.

[2]

Test statistic value: ______________ with p-value: ____________________

g. The researcher provided the following conclusion. Unfortunately, it is not quite appropriate. Edit it as ***briefly*** as possible to make it a correct conclusion.

[2]

The data do support that the average talk-time for these 20 new hydride batteries is greater than the average talk-time for these 25 cadmium batteries.

12. **NRG Supplements** – Many people use different energy supplements to help them stay awake and focused longer. On a timed brainteaser test, the computer will not allow a subject to go to the next question until they answer the question correctly. It is known that subjects who have taken no supplements prior to the test have an average completion time of 13 minutes. A researcher believes that if a subject takes an *NRG* supplement they will improve their time on the brainteaser test, on average. The researcher will test this theory at a 5% significance level.

a. State the hypotheses the researcher would like to test.

[3]

H_0: ____________________ H_a: ____________________

One-Sample Statistics				
	N	Mean	Std. Deviation	Std. Error Mean
comp_time	24	12.04	1.79	.365

One-Sample Test						
	Test Value = 13					
	t	df	Sig. (2-tailed)	Mean Difference	95% Confidence Interval of the Difference Lower	Upper
comp_time		23	.015	-.96	-1.72	-.20

b. The SPSS output above gives the result of the researcher's test. The test statistic value is missing from the output. Compute the appropriate test statistic value. Show your work.

[2]

Final answer: ____________________

c. Report the corresponding *p*-value for testing the hypotheses in part (a). Also provide a *well-labeled* sketch of the *p*-value.

[5]

Sketch:

p-value: __________

d. The *p*-value in part (c) tells us: (clearly circle one) [2]
   i. The probability that the null hypothesis is true.
   ii. The probability that the alternative hypothesis is true.
   iii. The largest level of significance, value of α, at which the null hypothesis can be rejected.
   iv. The smallest level of significance, value of α, at which the null hypothesis can be rejected.

e. Provide a brief well-written sentence regarding the conclusion (at the 5% level) in the context of the NRG supplement and the brainteaser test.

[2]

Statistics 250 Winter 2012 Exam 2

1. **Age at first marriage** – A researcher investigating changing trends in marriage wants to determine if the average age of first marriage is higher for men than for women. He takes a random sample of 30 recently married heterosexual couples from a list of newspaper wedding announcements and records the ages of the man and woman for each couple. An excerpt of the data is shown:

Couple	Male age	Female age	Difference (M – F)
1	27.3	27.6	-0.3
2	28.5	27.8	0.7
...			
30	30.7	32.4	-1.7
Mean	28.2	27.6	0.65
Standard Deviation	2.8	3.2	1.5

a. State the appropriate hypotheses to test the researcher's theory and complete the definition of the corresponding parameter. The research has decided to use a 5% significance level.

[3] **H_0:** ______________________ **H_a:** ______________________
where the parameter represents the population mean difference

____

Paired Samples Test

		Paired Differences							
					95% Confidence Interval of the Difference				
		Mean	Std. Deviation	Std. Error Mean	Lower	Upper	t	df	Sig. (2-tailed)
Pair 1	m - f	.65	1.50	.2746	.09	1.21	2.37	29	.024

b. Use the SPSS output to complete the statement by filling in the blank and circling the appropriate word:

[2] *The observed sample mean difference in ages is* ______________
standard errors (Circle one) ***ABOVE*** ***BELOW***
the hypothesized population mean difference in ages.

c. If there really is no difference in the average age of first marriage for men and women in the population of all such couples, what is the expected value of the test statistic?

[1] **Final answer** ______________

d. Provide a complete sketch of the *p*-value; include a label and values on the x axis, a full label for the distribution, the shaded area for the *p*-value, and the actual value of the *p*-value

[4]

***p*-value =** ______________

e. Circle the appropriate statistical decision: ***Reject H_0*** ***Fail to reject H_0***

[1]

f. What type of error could the researcher have made?

[1] (circle one) ***Type I Error*** ***Type II Error*** ***Cannot be determined***

2. **Caffeine Consumption** – Concerned about caffeine consumption by young adults, a doctor decides to investigate the caffeine consumption of college students and high school students. She randomly samples 72 college students (group 1) and 64 high school students (group 2) to investigate if the mean caffeine consumption (in mg) differs between college students and high school students. It has been a few years since she took her statistics course, so she uses Google to find a few commands in SPSS and obtains the output below:

Group Statistics

	student_typeh	N	Mean	Std. Deviation	Std. Error Mean
caffeine	college	**72**	**168.73**	**60.97**	**7.19**
	high school	**64**	**149.84**	**48.89**	**6.11**

Independent Samples Test

		Levene's Test for Equality of Variances		t-test for Equality of Means						
									95% Confidence Interval of the Difference	
		F	Sig.	t	df	Sig. (2-tailed)	Mean Difference	Std. Error Difference	Lower	Upper
caffeine	Equal variances assumed	.714	.400	1.98	134	.0501	18.89	9.56	-.01	37.79
	Equal variances not assumed			2.00	132.647	.0472	18.89	9.43	.23	37.55

a. The researcher wishes to estimate the difference in population mean caffeine consumption (college – high school) using a 95% confidence interval. She would like to use the pooled procedure. She notes that the two sample standard deviations are somewhat similar, but would like to be further assured by examining Levene's Test results.

i. Clearly state the null hypothesis for Levene's Test.

[1]

H_0: ____

ii. Explain how Levene's test results support her decision to pool (include numerical values).

[1]

b. In order to create the pooled confidence interval, an estimate of the common population standard deviation must be obtained. What is the value of this estimate? (Show all work).

[3]

Final answer ____________________

c. Consider the following statement written by the researcher in her report. It is not quite correct. Change ONE word in the statement to make it correct:

[1] *With 95% confidence, we would estimate the difference in sample mean caffeine consumption (College – High School) to be between -0.01 mg to 37.79 mg.*

3. Waiting at Wal-Mart – According to Wal-Mart, check-out time for their customers (during the 4 to 8 pm time period) has been uniformly distributed between 4 and 11 minutes. This rectangular shaped distribution has a mean of 7.5 minutes and a standard deviation of 2.02 minutes.

a. A summer intern working at Wal-Mart has been given the task to gather some data to assess if this model for check-out time is still reasonable. So he plans to take a random sample of 36 customers checking out between 4 and 8 pm, record each check-out time, and compute the average check-out time.

i. What is the approximate model (or distribution) for the average check-out time based on a random sample of 36 customers? Provide all details.

[3]

ii. What is the probability the average check-out time based on a random sample of 36 customers will be more than 8.5 minutes? Show all work.

[2]

Final answer ______________________

b. Suppose the intern was running out of time to complete the project and instead decided he would take a random sample of 9 customers. Explain how the distribution of the sample average in part (a) (i) above would be different by completing the following statements. [3 points]

i. The mean of the distribution would:
(circle one) ***increase*** ***decrease*** ***stay the same***

ii. The standard deviation of the distribution would:
(circle one) ***increase*** ***decrease*** ***stay the same***

iii. The shape of the distribution would:
(circle one) ***change*** ***stay the same***

---

4. Alpha – Environmentalists concerned about the impact of high-frequency radio transmissions on birds found that there was *no evidence* of a higher mortality rate among hatchlings in nests near cell towers. They based this conclusion on a test using $\alpha = 0.05$. Would they have made the same decision at $\alpha = 0.10$?

[2] Circle one: ***Yes*** ***No*** ***Can't Tell***
Briefly explain:

5. **Plastic Hardness** – Companies that manufacture plastic parts measure the hardness of their products in Brinell units. Different parts will have different minimal hardness criteria depending on the function of that particular part. Since measuring the hardness of a part (the Brinell score) involves *stressing* each part to the point of failure or breakage, companies often prefer to take a small random sample of parts.

   The population mean Brinell score for a particular plastic part in a new toy needs to be estimated using a 95% confidence interval. A random sample of 16 parts has been selected from last week's stable production run and the hardness (the Brinell score) has been obtained for each part.

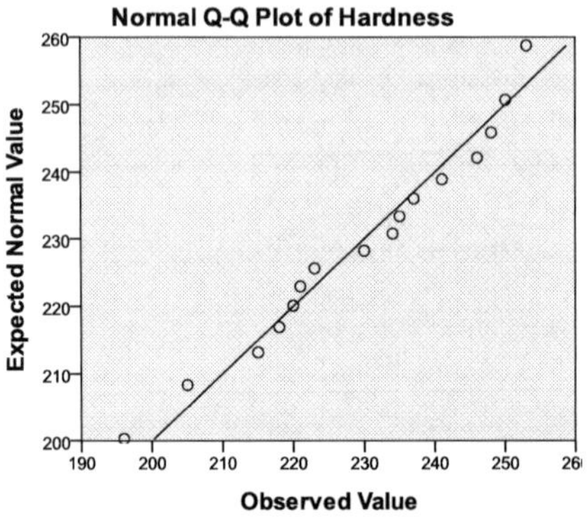

a. Before constructing the 95% confidence interval, the manager of the quality control group creates the graph you see at the right, along with a few summary measures. Clearly state the assumption that is assessed by looking at this graph.

[2]

PD-INEL

b. The sample mean Brinell score was 229.5 points. What was the standard deviation for these 16 observations? Show your work.

[2]

Final answer ____________________

c. Provide the 95% confidence interval for estimating the population mean Brinell score. Show your work.

[3]

Final answer: 95% Confidence Interval: ____________ **to** ____________

d. For this particular plastic part, the <u>minimal</u> level for acceptable hardness is 222 points. Based on the data, does it appear that this production run of plastic parts have acceptable hardness on average? Circle your answer and provide a brief explanation with numerical support for your answer.

[2] **Circle one:** **Yes** **No**

because . . .

6. **Herbs may ease Migraines –** A homeopathic preparation of ginger and feverfew offered some relief for migraine sufferers in a recent study. In the study, a group of 16 migraine sufferers treated themselves with the remedy as soon as they felt a migraine developing. Of the 16 subjects, 10 were pain free within three hours. The researcher used these results to test the claim (at a 5% significance level) that a majority of all migraine sufferers using this remedy would be pain free within three hours. Let p be the population proportion of all migraine sufferers using this remedy would be pain free within three hours.

a. State the appropriate hypotheses: **H_0:** ______________ **Ha:** ______________

[2]

b. Provide the observed test statistic value: ________________________

[1]

c. The p-value for the test is 0.227. What distribution was used to find this p-value? Provide all details.

[2]

Final answer ____

d. At a 5% significance level, the results

[1] (circle one) ***ARE*** ***ARE NOT*** statistically significant.

e. It has been recommended to repeat this study using a larger sample size. For the new study design (compared to the original study design), the probability of correctly *concluding* that a majority of all migraine sufferers using this remedy would be pain free within three hours ...

[2] Circle one: ***would increase*** ***would decrease*** ***would stay the same***

The statistical name for this probability is: ___________________________

f. Besides the original study design having a small sample of 16 subjects, what is another major issue regarding this study?

[1] Circle one: ***non-response bias*** ***response bias*** ***it is just an observational study***

7. **What is that p-value?** A school district was interested in assessing if the drop-out rate for their high school students is less than the national level of 10%, that is, testing H_o: $p = 0.10$ versus H_a: $p < 0.10$ at a 5% significance level. A large random sample of high school students from the district was selected and the sample proportion that dropped out was $\hat{p} = 0.12$.

[2] Which of the following is the only reasonable for the p-value?

Circle one: ***0.02*** ***0.05*** ***0.10*** ***0.12*** ***0.75***

8. Voter Appeal – In studying his campaign plans, Mr. Lehman wishes to learn if there is any difference in his voter appeal for male voters as opposed to female voters. His campaign manager takes two independent random samples and provides the following results:

	If the election were held today, would you vote for Mr. Lehman?		
Gender	**Yes**	**No**	**Total**
1 = Male	78	122	200
2 = Female	94	106	200

The campaign manager uses these results to test if there is a difference (at the 10% level) in the likelihood of voting for Mr. Lehman for the two populations of male voters and female voters, namely, to test: H_0: $p_1 = p_2$ versus H_a: $p_1 \neq p_2$

a. If the null hypothesis is true, there is just one common population proportion. Give an estimate of that common population proportion. Show all your work.

[2]

Final answer: ____________________________

b. The researcher remembers there is a data condition that needs checking before conducting the test. Provide that check (include numerical support).

[2]

c. The resulting test statistic value is -1.62. Provide the corresponding *p*-value. Show all work.

[2]

***p*-value:** ____________________________

d. Which of the following is the appropriate statistical decision and conclusion at the 5% level? **Check one.**

[2]

____ Reject H_0 and conclude there is not a significant difference in the likelihood of voting for Mr. Lehman for the two populations of male voters and female voters.

____ Reject H_0 and conclude there is a significant difference in the likelihood of voting for Mr. Lehman for the two populations of male voters and female voters.

____ Fail to reject H_0 and conclude there is not a significant difference in the likelihood of voting for Mr. Lehman for the two populations of male voters and female voters.

____ Fail to reject H_0 and conclude there is a significant difference in the likelihood of voting for Mr. Lehman for the two populations of male voters and female voters.

9. **On Time Orders –** A catalog sales company promises to deliver orders placed on the Internet within 3 days. Follow-up calls to a large random sample of recent customers gave a 95% confidence interval for the proportion of all orders that arrive on time to be (82%, 92%).

a. What proportion of sampled orders arrived on time?

[2] Circle one: ***82%*** ***92%*** ***87%*** ***10%*** ***5%*** ***Can't be determined since we do not know the sample size n***

b. The margin of error of this interval is:

[2] Circle one: ***82%*** ***92%*** ***87%*** ***10%*** ***5%*** ***Can't be determined since we do not know the sample size n***

c. The researcher felt that the confidence interval was too wide to provide a precise estimate of the population proportion. What could the researcher have done to produce a narrower (more precise) confidence interval estimate of the population proportion of all orders arriving on time? Circle your answer.

[2] ***i. Increase the confidence level to 99%.***

ii. Increase the sample size.

iii. Both i and ii.

iv. None of the above.

---

10. **Name that Scenario –** Below are a number of research problems. For each, determine if the most appropriate statistical analysis technique is conducting a hypothesis test (HT) or constructing a confidence interval (CI). Then also give the symbol for the corresponding parameter that is of interest. [6 points]

a. Police set up an auto checkpoint at which drivers are stopped and their cards inspected for safety problems. They want to estimate the percentage of all cars on the road that may be unsafe.

Circle one: HT CI Parameter of interest: ________________

b. A company has recently started an exercise break program for its workers to see if it will improve job satisfaction on average, as measured by a questionnaire with higher scores indicating more satisfied. Scores for 10 randomly selected workers before and after the implementation of the exercise program were recorded.

Circle one: HT CI Parameter of interest: ________________

c. Researcher Colinda would like to determine if the proportion of all adults favoring the death penalty is the higher than that for all teenagers. She asks a random sample of adults (group 1) whether or not they favor the death penalty. She also asks an independent random sample of teens (group 2) whether or not they favor the death penalty.

Circle one: HT CI Parameter of interest: ________________

11. Carbon Monoxide in LA – The Environmental Protection Agency (EPA) conducted a test to assess if the population average carbon monoxide level in the downtown Los Angeles area is higher than 4.9 parts per million (ppm).

The hypotheses tested were: H_0: $\mu = 4.9$ versus H_a: $\mu > 4.9$ and a 5% significance level was used. The results from a random sample of 25 carbon monoxide readings gave a sample mean of 5.40, a sample standard deviation of 1.03, a test statistic value of 2.43 with a corresponding *p*-value of 0.011.

Below are *possible* statements for including in the EPA report. You have been asked to determine which statements are appropriate to include (True) and which are not appropriate to include (False). Clearly circle your answer for each statement. [1 point each]

a. The value of 5% represents the tolerable risk of concluding the population average carbon monoxide level is greater than 4.9 ppm, when in fact, it is not.	**True** **False**
b. The power of the test is the probability that the EPA correctly concludes the population average carbon monoxide level is greater than 4.9 ppm.	**True** **False**
c. The standard normal distribution was used to calculate the *p*-value of 0.011.	**True** **False**
d. The *p*-value of 0.011 is the probability that the null hypothesis is true.	**True** **False**
e. Based on the data, the carbon monoxide level in all areas in downtown Los Angeles is greater than 4.9.	**True** **False**

12. Memory Experiment – A researcher conducts an experiment on human memory and recruits 10 people to participate in her study. She performs the experiment and analyzes the results. She obtains a *p*-value of 0.13. Which of the following is a reasonable interpretation of her results? Her significance level was 10%.

[2] Circle one:

i. ***This proves that her experimental treatment has no effect on memory.***

ii. ***She should reject the null hypothesis.***

iii. ***There could be a treatment effect, but the sample size was too small to detect it.***

Final Exam Questions

Statistics 250 Fall 2011 Final Exam

1. Does the distribution of lunch drink preference differ for teenagers versus adults? A random sample of 400 teens and an independent random sample of 600 adults were surveyed to determine the type of drink they prefer to drink with their lunch. The table below provided the results. **[12 points]**

Preferred Lunch Drink	Teens	Adults	Total
Coffee	105	140	**245**
Tea	100	150	**250**
Soft Drink	220	180	**400**
Other	50	55	**105**
Total	**400**	**600**	**1000**

a. To assess if the distribution of lunch drink preference differ for the teenager population versus the adult population, the appropriate test to conduct would be:

[2] Circle one: **goodness-of-fit** **homogeneity** **independence**

b. If the distribution of lunch drink preference is the same for the two populations, how many teens would you expect to prefer coffee?

[2]

Final answer = ____________________

c. If the distribution of lunch drink preference is the same for the two populations, what is the expected value of the test statistic?

[1]

Final answer = ____________________

d. The data were entered into SPSS and part of the output is given below which include the *p*-value.

Chi-Square Tests

	Value	df	Asymp. Sig.
Pearson Chi-Square	16.78	3	0.0008

[3] Provide a well-labeled sketch showing this *p*-value →

e. At a 5% significance level, it appears that the distribution of lunch drink preference...

[1] (circle one) ***is*** ***is not*** the same for the teenager population versus the adult population.

2. A psychiatrist wants to know how different therapies affect the average time per month (measured in days) a patient experiences symptoms of some psychological disorder. Based on his clinical experience, there are three types of therapies to consider: medical, behavioral, and a combination of the two. To investigate if there is a difference, he randomly assigns 7 patients to receive medical therapy, 5 patients to receive behavioral therapy, and the remaining 8 patients to receive the combination therapy. **[13 points]**

a. The psychiatrist's study is an

[1] (circle one): ***OBSERVATIONAL STUDY*** ***EXPERIMENT***

b. As the psychiatrist is comparing three therapies with respect to the length of time each month a patient experiences symptoms (in days), he performs an ANOVA. Partial output is given below. Fill in the blanks to complete the ANOVA table.

[4]

Days with symptoms **ANOVA**

	Sum of Squares	df	Mean Square	F	Sig.
Between Groups	**(i)**	**(ii)**	24.626	**(iv)**	.000008
Within Groups	16.500	**17**	.971		
Total	65.753	**(iii)**			

(i) = ____________________

(ii) = ____________________

(iii) = ____________________

(iv) = ____________________

c. One of the assumptions to perform ANOVA is that the population variances are equal. If they are the same, then they equal some common value. Provide an estimate of this common population variance.

[2]

Final answer = ____________________

d. Based on the ANOVA table, at a 10% level, what is the correct statistical decision and conclusion?

[1]

- Reject H_0, conclude there is sufficient evidence that at least one population mean time with symptoms is different.
- Reject H_0, conclude there is sufficient evidence that the population mean times with symptoms are equal.
- Fail to reject H_0, conclude there is sufficient evidence that at least one population mean time with symptoms is different.
- Fail to reject H_0, conclude there is insufficient evidence that at least one population mean time with symptoms is different.

e. Tukey's multiple comparison tests were conducted and the corresponding SPSS output is provided below. Do the Medical Therapy and the Behavioral Therapy appear to be different in terms of the population mean length of time with symptoms (per month) at the 5% level?

Days_with_symptoms
Tukey HSD

Multiple Comparisons

(I) Therapy_type	(J) Therapy_type	Mean Difference (I-J)	95% Confidence Interval Lower Bound	95% Confidence Interval Upper Bound
Medical	Behavioral	.949	-.531	2.429
	Combination	3.512	2.204	4.820
Behavioral	Medical	-.949	-2.429	.531
	Combination	2.563	1.122	4.004
Combination	Medical	-3.512	-4.820	-2.204
	Behavioral	-2.563	-4.004	-1.122

[2] *YES* *NO*

Briefly explain using numerical support:

f. Provide a 99% confidence interval for the population mean time with symptoms for the medical therapy group alone, in which the 7 patients reported an average of 9.17 days with symptoms each month.

[3]

Final answer: (________________ , ________________)

3. **Short Concept Questions – 2 points each**, clearly circle your one answer for each part. **[8 points]**

a. A confidence interval is a range of reasonable values used to estimate:

   i. a population parameter.
   ii. a sample statistic.
   iii. the sampling distribution of the sample statistic.
   iv. the shape of the sampling distribution.

b. Which of the following does NOT need to be known in order to compute the p-value?

   i. The level of significance.
   ii. The value of the test statistic.
   iii. The direction of extreme (usually in the alternative hypothesis).
   iv. The distribution of the test statistic under the null.

c. A 95% confidence interval is computed to estimate the mean household income for a city. Which of the following values will definitely be within the limits of this confidence interval?

   i. The population mean
   ii. The sample mean
   iii. The standard deviation of the sample mean
   iv. None of the above

d. A researcher conducts an experiment on human memory and recruits 15 people to participate in her study. She performs the experiment and analyzes the results. She obtains a p-value of 0.12. Which of the following is a reasonable interpretation of her results? Her significance level was 10%.

   i. She should reject the null hypothesis.
   ii. This proves that her experimental treatment has no effect on memory.
   iii. There could be a treatment effect, but the sample size was too small to detect it.

---

4. We are interested in conducting a study to estimate the proportion of all eligible voters who would vote for the incumbent governor. What is the minimum sample size needed to estimate this proportion with 90% confidence and a margin of error of (no more than) 5%? Show your work. **[3 points]**

Final answer: ______________________________

5. During the last two weeks of December, airlines in the U.S. transport many people around the country. Unfortunately, these weeks often coincide with snowstorms that cause flight cancellations and delays. A consumer advocacy group wants to know if budget airlines have ***longer*** delays than standard commercial airlines. They take a random sample of 20 customers who flew on a budget airlines and a random sample of 20 customers who flew on a standard commercial airlines during the two week period at the end of December and asked them how long their flight was delayed (in hours, with 0 would be entered if the flight had no delays). The analyst from the consumer advocacy group working on this project entered the data into SPSS and produces the following two sets of outputs. **[11 points]**

Paired Samples Test

	Paired Differences					t	df	Sig. (2-tailed)
	Mean	Std. Deviation	Std. Error Mean	95% Confidence Interval of the Difference: Lower	Upper			
Pair Commercial - Budget	-.317	2.25	.5031	-1.370	.736	-.630	19	.536

Independent Samples Test

1 = Commercial 2 = Budget		Levene's Test for Equality of Variances		t-test for Equality of Means						
		F	Sig.	t	df	Sig. (2-tailed)	Mean Difference	Std. Error Difference	95% Confidence Interval of the Difference: Lower	Upper
Delay	Equal variances assumed	3.801	.059	-.573	38	.5700	-.317	.553	-1.436	.802
	Equal variances not assumed			-.573	35.296	.5703	-.317	.553	-1.439	.805

a. Which output should be used?
[2] (circle one) ***Paired Samples Test*** ***Independent Samples Test***

b. Write the necessary hypotheses for the investigation.

[3] H0: ____________________ Ha: ____________________

c. Provide the appropriate test statistic and corresponding *p*-value.
[3]
Test statistic: ______________ *p*-value: ______________

d. Fill in the blank to correctly complete the sentence:
If this study were repeated many times and delays for budget airlines are no longer than those for standard commercial airlines, the analyst could expect to obtain a test statistic value as extreme
[1] or more extreme that observed in part (c) ______________% of the time.

e. At a 5% significance level, the data (circle one) ***ARE*** ***ARE NOT*** statistically significant.
[1]

f. The research team decides 20 is a small sample size so they conduct a follow up study with more people. In the follow up study, the probability of concluding delays are longer for budget airlines when delays really are longer for budget airlines would be __________ than the same probability for the original study.
[1] Circle one: ***SMALLER*** ***LARGER*** ***THE SAME*** ***CAN'T TELL***

6. A sociologist is studying how household chores are divided up for married couples. She surveys a random sample of 18 married couples who live together and records how many hours each spends on household chores (cleaning, cooking, errands, etc.) each week.

   The resulting **99% confidence interval** for the population mean difference in time spent on household chores (wife time less husband time) was found to be (1 hour, 6 hours). **[8 points]**

a. The sociologist has drafted a number of statements that she would like to include in her report.

   Which of the following would be correct statements that could be included? Circle all that are correct.

   - With 99% confidence, wives are estimated on average to spend 1 to 6 hours more time each week on household as compared to their husbands.
   - For the sample of 18 married couples, the mean difference in time spent on chores was 3.5 hours.
   - If this study were repeated many times, we would expect 99% of the resulting intervals to contain the true population mean difference in time spent on household chores.
   - If this study were repeated many times, 99% of the time we would expect the true population mean difference in time spent on household chores to fall in the interval (1, 6).
   - This confidence interval requires the distribution of time spent on household chores by the wives to be normally distributed.
   - At a 1% significance level there is sufficient evidence to say married women who live with their husbands spend more time than their husbands on household chores each week, on average (for the population of all couples represented by this sample).
   - A 90% confidence interval made with the same data would be narrower.

b. It has been suggested by another sociologist that the age of the individuals in this study might impact how many hours husbands and wives spent performing household chores.

   In this case, age would be an example of a/an

   ______________________________ variable.

7. **Muscle Mass –** A person's muscle mass is expected to generally decrease with age. Using a 5% significance level, we are to explore this relationship for women. A nutritionist randomly selected 60 women ranging from 40 to 79 years old and measured the muscle mass of each in pounds. **[18 points]**

a. For this study, the variable ***age*** plays the role of the ______________________________ variable.

[1]

b. The scatterplot of the data is provided.
Give a complete interpretation of this scatterplot.

$\overline{x}$ = 59.98 years $\overline{y}$ = 84.43 pounds

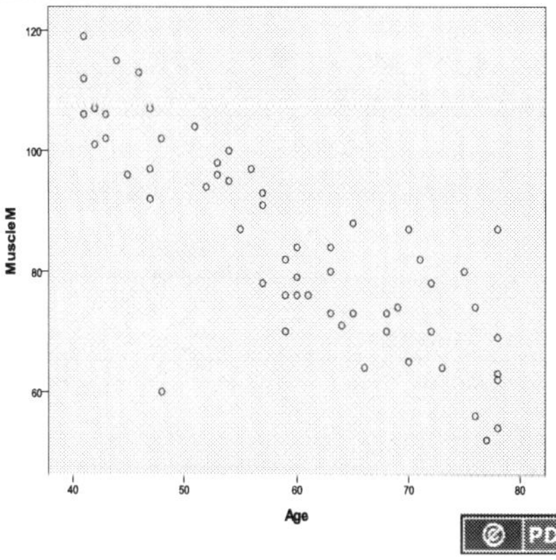

After a comprehensive review of the data, it was decided that all data points were valid. Various SPSS output is included to help you answer the following questions.

Descriptive Statistics

	N	Minimum	Maximum	Mean	Std. Deviation
Age	60	41	78	60.08	11.707
MuscleM	60	52	119	84.43	16.498

Model Summary

Model	R	R Square	Adjusted R Square	Std. Error of the Estimate
1	.818	.668	.663	9.583

ANOVA

Model		Sum of Squares	df	Mean Square	F	Sig.
1	Regression	10732.747	1	10732.747	116.880	.00000
	Residual	5325.986	58	91.827		
	Total	16058.733	59			

Coefficients

Model		Unstandardized Coefficients		Standardized Coefficients	t	Sig.
		B	Std. Error	Beta		
1	(Constant)	153.012	6.463		23.676	.00000
	Age	-1.143	.106	-.818	-10.811	.00000

c. Subject #10 was 48 years old with a muscle mass of 102 pounds. Compute her predicted muscle mass and then the corresponding residual (include your units).

[3] Predicted value = ______________________

and residual = ______________________________

d. The correlation between age and muscle mass is: ____________________

[2]

e. The nutritionist would like to assess if muscle mass generally decreases with age. State the appropriate hypothesis to be tested:

[4] H_0: ____________________ versus H_a: ________________________

Give the test statistic value that would be used to test these hypotheses:

____________________.

The p-value is less than 0.00001. Give the complete distribution that was used for finding this p-value: ______________________________

f. Before we assess if there is a significant indirect linear relationship between age and body mass for 40 to 79 year old women, we should check assumptions. Use the plots to answer the bulleted questions.

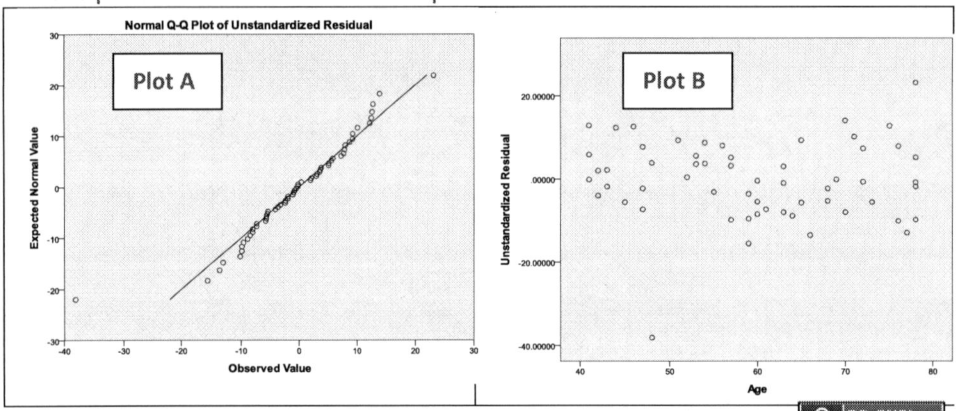

- Plot A on the left is used to assess the normality of the true errors, the true differences $y - (\beta_0 + \beta_1 x)$ for all possible pairs (x,y) for this population of women. ***TRUE*** ***FALSE***

[1]

- Plot B on the right is used to further check a condition. Complete the statement below to correctly state that condition.

[1] ***The variability of the true errors ...***

____

g. The researcher was going to use the regression equation to estimate the muscle mass for a 35 year old woman. Briefly explain why this should this be avoided.

[1]

h. The first interval estimate made was a 95% confidence interval for the mean muscle-mass for all 48-year-old women which went from 94.6 pounds to 101.7 pounds. The second confidence interval to be made will be the 95% confidence interval for the mean muscle-mass for all 57-year-old women. Without computations, briefly explain why this second confidence interval will be narrower than the first.

[2]

---

8. Five years ago, the student body of a local college consisted of 30% freshmen, 24% sophomores, 26% juniors, and 20% seniors. A random sample of 290 students taken from this current year's student body showed the following number of students in each classification. We are interested in determining whether or not the model for student classification this year has significantly changed from that of 5 years ago. **[7 points]**

Class	**Freshman**	**Sophomore**	**Junior**	**Senior**
Number	67	72	80	71

a. State the appropriate null hypothesis.

[2] H_0: ____

b. If the model for student classification this year has not changed from that of 5 years ago, how many of the selected students would have expected to be Sophomores? Show all work.

[2]

Final answer: ____________________

c. The observed test statistic value is 7.88. Assuming the model has not changed from that for 5 years ago, about how many standard deviations is the observed test statistic value from the expected value? Show your work.

[3]

Final answer: ____________________________

9. **Law School?** The LSAT (Law School Admissions Test) normalizes the scores so that each student's overall score ranges from 120 to 180. The scores for all students taking the exam are treated as continuous and modeled using a normal distribution with a mean of 150 and a standard deviation of 10. *Show all work when answer the following questions.* **[7 points]**

a. What is the probability a randomly selected student score is at least 162 points?
[2]

Final answer = ____________________

b. If you knew a particular student scored above the mean of 150 points, what is the probability that they actually scored at least 162 points?
[2]

Final answer = ____________________

c. Consider a random sample of 10 students applying for law school who have taken the LSAT. What is the probability that exactly two of the 10 students scored at least 162 points?
[3]

Final answer = ____________________

---

10. Tent Stakes – Big John makes tent stakes for every kind of tent. The machinery in his factory is somewhat old and the length of the test stakes often have some variability. Big John's nephew (who studied statistics at UM) determined that the model for lengths of the 40-inch tent stakes were uniformly distributed over the range of 38 to 42 inches and the model for the lengths of the 43-inch tent stakes were uniformly distributed over the range of 40 to 46 inches. Below are the sketches of these two models. **[4 points]**

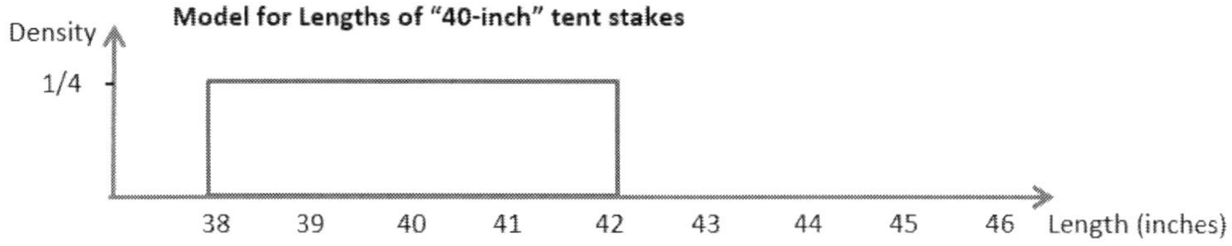

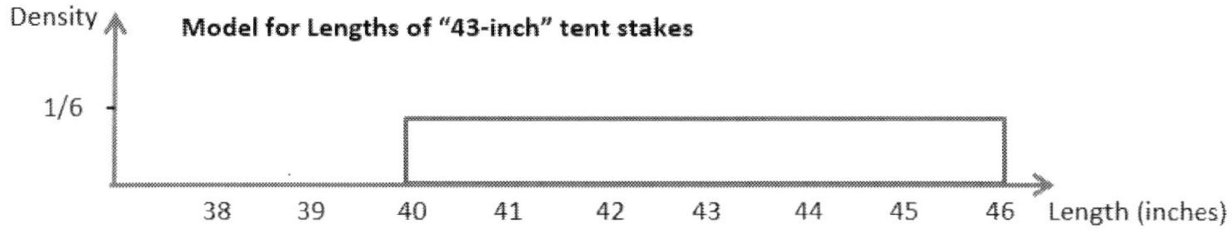

While loading boxes of 40-inch tent stakes and boxes of 43-inch tent stakes onto the truck, the crew found a box on the factory floor without its label. The foreman's advice was to pick one tent stake from the box at random and **if the length was 41 inches or longer, put a label of "43-inch tent stakes" on that box.** So the crew will be picking between the following hypotheses:

H_0: The box is filled with "40-inch" tent stakes

versus H_a: The box is filled with "43-inch" tent stakes

using the decision rule: **Reject H_0 if the length of the randomly selected tent stake is 41 inches or longer.**

a. For this decision rule, find the level of significance, that is, compute α.
[2]

Final answer = ____________________

b. For this decision rule, compute the statistical power.
[2]

Final answer = ____________________

11. The Web site www.twiigs.com allows you to vote on polls that interest you or to post one of your own. Once you have found a poll of interest, you just click on "Vote," and your response becomes part of the sample. One of the questions was "How many times have you been pulled over by the police?" Of the 780 people who responded, 70% said "1-5 times."
You conclude the results tell us little about the population of all adults because they are subject to bias.

[1] The name of the ***primary*** bias here is
____________________________________ bias.

12. **Name that Scenario** – For each research problem below, determine if the appropriate method to address the problem would be to make a **confidence interval (CI)** or to **conduct a test of hypotheses (HT)**.

If a CI, then provide the notation for the corresponding parameter the CI is for.

If a HT, then clearly state the appropriate null and alternative hypotheses to be tested. The last scenario has one additional question, so be sure to answer it too. **[11 points]**

a. Researchers Emilio and Emily want to learn about the proportion of a certain type of tree growing in a national forest that suffers from a disease. They take a representative sample of 200 of the trees from the forest and find that 15 of the 200 sampled trees have the disease.

CI for ____________ ***OR*** **H_0:** ______________ **versus H_a:** ______________

b. The manager of the Chatta Department store at the North Mall is interested in estimating the difference between the mean credit purchase of customers using the store's credit card versus the mean credit purchase for those customers using a national major credit card.

CI for ____________ ***OR*** **H_0:** ______________ **versus H_a:** ______________

c. A random sample of twenty-five college women was asked for their own heights and their mothers' heights. The researchers wanted to know whether college women differ in height on average compared to that of their mothers.

CI for ____________ ***OR*** **H_0:** ______________ **versus H_a:** ______________

d. The average monthly rent for a one-bedroom apartment in Chattanooga has been $700. With the downturn in the real estate market, a study will be conducted to assess if there has been a decrease in the average rental cost.

CI for ____________ ***OR*** **H_0:** ______________ **versus H_a:** ______________

Clearly define the parameter of interest:

The parameter __________________ represents __________________________

____

Statistics 250 Winter 2012 Final Exam

1. Favorite Subject Area – A random sample of 300 students were asked which of three areas he or she preferred taking liberal arts courses, 1 = Math/Science, 2 = Social Science, or 3 = Humanities. The data were further broken down to compare the results by gender. The results are provided in the following table.

	Preferred Area for Liberal Arts Courses			
Gender	**1 = Math/Science**	**2 = Social Science**	**3 = Humanities**	**Total**
Male	37	41	44	122
Female	35	72	71	178
Total	72	113	115	300

a. The researcher set out to assess if the distribution of the preferred area for liberal arts courses is the same for the two populations of all male students versus all female students. What is the appropriate chi-square test to perform?

[1] Circle one: **goodness-of-fit** **homogeneity** **independence**

b. Complete the sentence by providing the appropriate percentages:

[2] The results show that overall 24% of the surveyed students preferred the Math/Science area; however focusing on just the male students, the *percent* preferring the Math/Science area was ____________________________
and focusing on just the female students, the *percent* preferring the Math/Science area was __________________________________.

c. If the distribution of the preferred area for liberal arts courses is the same for the two populations of all male students versus all female students, how many male students would you have *expected* to prefer the Math/Science area? Show your work.

[2]

Final answer: ________________________

d. The data were entered into SPSS and part of the output is given below which include the *p*-value.

Chi-Square Tests

	Value	**df**	**Asymp. Sig.**
Pearson Chi-Square	4.604	2	0.1001

[3] Provide a well-labeled sketch showing this *p*-value →

e. At a 5% significance level, it appears that the distribution of the preferred area for liberal arts courses ...

[1] (circle one) ***is*** ***is not*** the same for the two populations of all male students versus all female students.

2. **Raffles for Prizes –** For a fundraiser, a local school will raffle off two donated prizes – a new TV and a new iPhone. The 200 attendees at the fundraiser were allowed to purchase a ticket for one or the other or both raffles if they wished. By the end of the night, 40% of the attendees purchased a ticket to try to win the TV, 75% purchased a ticket to try to win the iPhone, and 30% purchased a ticket for both raffles.

a. What percent of the attendees did not purchase any ticket for either of the prizes?

[2]

Final answer: ______________________

b. Given a randomly selected attendee did purchase a ticket to try to win an iPhone, what is the probability he or she also purchased a ticket to try to win the TV? Show your work.

[2]

Final answer: ______________________

c. Circle all that apply about the events I = {purchased a ticket to win iPhone} and T = {purchased a ticket to win TV}:

[2]

Independent

Mutually exclusive

Complementary

None apply

3. **Rain, Rain, Go Away** – Suppose the annual rainfall in Ann Arbor follows a uniform distribution from 15 to 55 inches.

a. Provide a complete sketch of the distribution of the annual rainfall in Ann Arbor. Provide values and labels for both axes.

[3]

b. A student studying meteorology believes the rainfall last year, which was 50 inches, was quite unlikely. Find the probability that the annual rainfall for a year would be as much as 50 inches or more.

[1]

Final answer: ______________________

c. Suppose it can be assumed that the annual rainfall is independent from year to year. What is the probability that at most one of the next 10 years will have an annual rainfall of 50 inches or more? Show your work.

[3]

Final answer: ______________________

4. **Intramurals** – The student government designed a survey to gather opinions of fellow students at Albion College regarding various line items in the proposed college budget for next year. A random sample of 40 students was selected. One question on the survey asked whether they felt the amount being budgeted for intramural sports was an appropriate amount, and 45% responded 'yes'.

a. Upon reading a draft of the survey results which reported this sample proportion $\hat{p}$ of 0.45, the student government leader requested that the standard deviation of $\hat{p}$ also be provided. A student taking Statistics 250 responded to his request saying that the standard deviation of $\hat{p}$ cannot be reported as *p* is unknown.

[1] Is this comment by the Stats 250 student correct?

(circle one) ***Yes*** ***No***

b. The corresponding 90% confidence interval to estimate the proportion of all Albion College students that think this budgeted amount is appropriate was found to go from 32.1% to 57.9%. Does is appear there is statistically significant support for a "minority"?

[2]

(circle one) ***Yes*** ***No***

because . . .

c. The 95% confidence interval would be...

[1] (circle one) **wider than** **narrower than** **the same length as**

the 90% confidence interval in part (b).

d. The Director of Athletes has asked the student government to repeat this survey taking a larger random sample of Albion students. How many students should be surveyed (at a minimum) in order for the width of the 90% confidence interval be at most 0.05 (or 5%)? Show all work.

[2]

Final answer: ____________________

5. **Student Interpretations –** Judy is working on a statistics project using data on pulse rates (in bpm = beats per minute) collected from a random sample of 100 students from her college. Her 99% confidence interval for the population mean pulse rate is (66 bpm, 72 bpm). She provides you the following list of possible statements and interpretations to include in her report. Circle the statements that are correct and can be considered for inclusion.

[4]

- We are confident that 99% of all students at this college will have pulse rates between 66 and 72 beats per minute.
- We are 99% confident that the sample mean pulse rate will be between 66 and 72 beats per minute.
- We would estimate, with 99% confidence, the mean pulse rate for all students at this college to be 69 beats per minute with a margin of error of 3 beats per minute.
- We would expect 99% of all possible random samples of this same size to result in a confidence interval that would contain the population mean pulse rate for all students at this college.

---

6. **Test scores and GPA** – A college administrator wished to determine if standardized test scores are a good linear predictor of college GPA for students at her small college. She took a random sample of 30 such students and recorded their (end of first college year) GPA and ACT score from their records. She first examined the scatterplot to see that a general linear relationship was initially reasonable. She then performed a full regression analysis and some of the output from SPSS is given below.

ANOVA

Model		Sum of Squares	df	Mean Square	F	Sig.
1	Regression	2.115	1	2.115	10.729	.003
	Residual	5.521	28	.197		
	Total	7.636	29			

Coefficients

Model		Unstandardized Coefficients		Standardized Coefficients	t	Sig.
		B	Std. Error	Beta		
1	(Constant)	1.501	.552		2.718	.011
	ACT	.067	.020	.526	3.275	.003

PD-INEL

a. In this regression analysis, the variable **first year college GPA** plays the role of the ________________ variable.

[1] Circle one: ***response*** ***explanatory*** ***confounding***

b. Provide the equation of the least squares regression line.

[2]

Final answer: ____

c. One of the students in the data set had an ACT score of 32 and a first year college GPA of 3.92. What is the value of the residual for this student? Show all work.

[2]

Final answer: ____________________

d. The administrator wanted to test if there is a significant *positive* linear relationship between ACT scores and GPA. Conduct this test at a 5% level by providing the appropriate hypotheses, test statistic, and *p*-value.

[4]

H_0: ____________________ versus H_a: ____________________

Test Statistic Value: ____________________ *p*-value: ____________________

e. Based on the results in part (d), the administrator can conclude

[1] (circle one) ***THERE IS*** ***THERE IS NOT*** a significant positive linear relationship between ACT scores and first year college GPA.

f. Complete the sentence below (show your work):

[2]

Based on this analysis, ____________________ % of the variability in first year college GPAs is ***not*** accounted for by its linear relationship with ACT scores.

g. The administrator obtained a 95% prediction interval for the GPA of an incoming student with an ACT score of 30. The prediction interval was (2.62, 4.41). Which of the following intervals must be the 95% confidence interval for the mean GPA of all students with a score of 30 on their ACT?

[2] Circle one: **(2.53, 4.57)** **(3.31, 3.72)** **(3.02, 3.53)**

h. Consider the following plot created by the administrator.

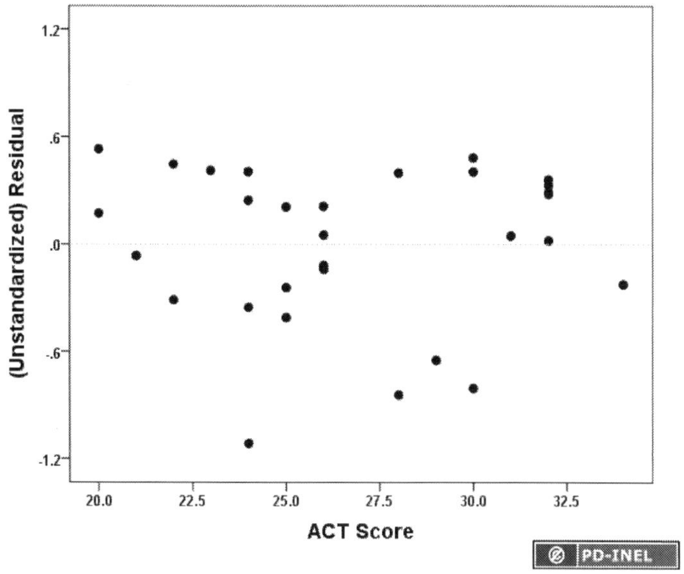

[2] This plot is used to check which assumption in regression analysis (circle your answer):

- the ACT scores are normally distributed.
- the population of errors have constant variance and no obvious patterns.
- the unstandardized residuals have constant variance and no obvious patterns.
- the ACT scores are randomly distributed.

i. Complete the following sentence:
[1] If the administrator had asked the random sample of students to self-report their first year college GPA and ACT scores, then she may have encountered some bias in her results due to ________________________ bias.

7. **Magnet vs Non-Magnet Hospitals** – Nursing Magazine reported the following model for staffing status at magnet hospitals nationwide. (A magnet hospital is one that successfully attracts and retains nurses.)

Staffing Status	**National "Magnet" %**
1 = Desperately short (patient care as suffered)	12%
2 = Short, but patient care hasn't suffered	32%
3 = Adequate	38%
4 = More than adequate	18%

The Arkansas Nursing Association conducted a survey of $n = 500$ nurses chosen at random from non-magnet hospitals in the state. The results of this survey are summarized below.

Staffing Status	**Arkansas Sample Results**
1 = Desperately short (patient care as suffered)	165
2 = Short, but patient care hasn't suffered	140
3 = Adequate	125
4 = More than adequate	70

a. One goal of the Arkansas study is to assess if the model for staffing status for non-magnet hospital nurses in Arkansas follows the model for magnet hospitals nationwide. Provide the corresponding null hypothesis.

[2]

H_0: ______________________________

b. In the Arkansas study, we observed 70 nurses that classified the staffing status to be "more than adequate". How many nurses from this sample would we have *expected* to answer "more than adequate" if the model for staffing status for non-magnet hospital nurses in Arkansas follows the model for magnet hospitals nationwide? Show your work.

[2]

Final answer: ______________

c. The observed test statistic value turns out to be very large and the corresponding *p*-value is very small. Assuming the model of staffing status for non-magnet hospital nurses in Arkansas really does follow the model for magnet hospitals nationwide, what distribution would have been used to find the corresponding *p*-value?

[2]

Final answer: ____________________

d. If the model of staffing status for non-magnet hospital nurses in Arkansas really does follow the model for magnet hospitals nationwide is true, what would have been the *expected value* of that test statistic?

[2]

Final answer: ______________

8. **Matching *p*-values** – Match the four *p*-values with the appropriate conclusions.

[4] *p*-values: A = 0.0080 B = 0.0320 C = 0.0621 D = 0.2800

Conclusion	*p*-value (circle your answer)			
The evidence against the null hypothesis is significant, but only at the 10% level.	A	B	C	D
The evidence against the null hypothesis is very strong.	A	B	C	D
There is not enough evidence to reject the null hypothesis, even at the 10% level.	A	B	C	D
The evidence is significant at a 5% level, but not at a 1% level.	A	B	C	D

9. **Yoga Course** – A yoga instructor is working on a project in which some high school (HS) students participated in a yoga course. The length of time (seconds) that a student could correctly hold a certain yoga pose was collected at the start of the course and then at the end. The instructor wanted to learn if their yoga skills for this pose improved on average (where improvement implies they could hold the correct pose longer). For each of the 10 students in the study, improvement was measured as **Time at End** *less* **Time at Start**. Some SPSS output is provided below.

Paired Samples Statistics

		Mean	N	Std. Deviation	Std. Error Mean
Pair 1	Time at End	20.70	10	10.863	3.435
	Time at Start	15.10	10	6.773	2.142

Paired Samples Test

		Paired Differences					t	df	Sig. (2-tailed)
					95% Confidence Interval of the Difference				
		Mean	Std. Deviation	Std. Error Mean	Lower	Upper			
Pair 1	Time at End - Time at Start	5.600	6.450	2.040	.986	10.214	2.746	9	.023

a. The parameter of interest is μ_D, the population mean difference in length of time they can hold the pose (time at end – time at start) for the population of all high school students. Keeping in mind that a score of 0 indicates no improvement, what type of differences would indicate that the yoga course was effective?

[1] Circle one: ***positive*** ***negative***

b. State the hypotheses for assessing if yoga skills for this pose improved on average.

[2]

H_0: ______________________ versus H_a: ______________________

c. Do the data support the yoga skills for this pose improved on average? Circle your answer to this question, then provide the test statistic value and the corresponding *p*-value for support. Use a 5% significance level.

[3] Circle one: ***Yes*** ***No***

Test statistic: ___________________

p-value (for the hypotheses in part b): ____________________________

d. The yoga instructor remembers from her statistics course that there are "conditions" for any test, to ensure its integrity. She believes the 10 high school students that participated in the study are representative of the population. But there is an additional "condition". Clearly state that condition.

[2]

The condition: ____

____

---

10. Significance Level and Type I Error – A new drug will be tested with potentially dangerous side effects to see if it is significantly better than the drug currently in use. The hypotheses (in words) are given as H_0: The new drug is not better than the current drug versus H_a: The new drug is better than the current drug. If the new drug is found to be significantly more effective, it will be prescribed to millions of people.

a. Briefly describe what it means in this particular situation to make a Type I error.

[2]

b. Would it make more sense to use a large significance level (10%) or a small significance level (1%)?

[1] Circle one: ***10%*** ***1%***

11. **Sulfur Dioxide Concentration –** A study was conducted to compare the mean sulfur dioxide concentrations for three cities. Independent random samples, each of size 3, were obtained from each city and the concentration levels were measured. The researcher wishes to perform an ANOVA to assess if the population mean concentration levels for the three cities are the same using a 10% significance level. Part of the ANOVA table for testing this hypothesis is provided below.

ANOVA

sulfur dioxide concentration

	Sum of Squares	df	Mean Square	F	Sig.
Between Groups	24.0	2	12.0	?	.064
Within Groups	16.0	6	2.7		
Total	40.0	8			

a. Before conducting the test to assess if the population mean concentration levels for the three cities are the same, several plots of the sulfur dioxide concentration data should be made to help verify some of the assumptions. The student assistant working on this study created a qq-plot of the 9 concentration measurements as shown at the right.

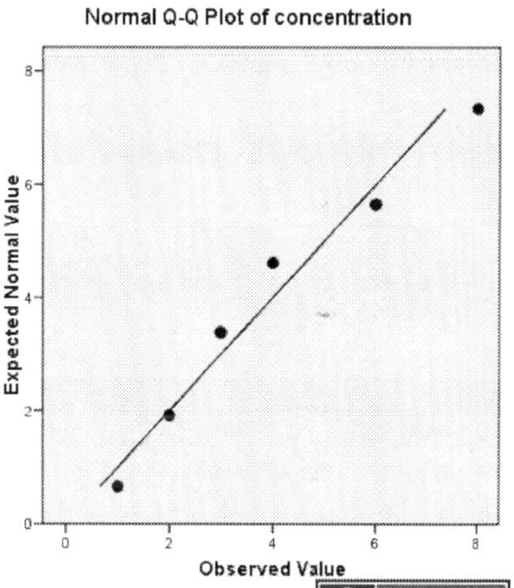

[2]

Is this the appropriate qq-plot to examine?
Circle one: ***Yes*** ***No***

If yes, *briefly* explain which assumption is being checked and whether the assumption seems reasonable.
If no, *briefly* explain why this is not an appropriate plot.

b. The test statistic for testing the hypothesis that the population mean concentration levels for the three cities are the same is missing. Compute that missing value.

[1]

Final Answer: ______________________

c. What distribution was used to find the reported *p*-value provided in the output above? Be specific.

[2]

Final Answer: ____

d. Consider the following four statements. Clearly circle *all* that are correct regarding this ANOVA study.

[3]

i. If this study were repeated many times, the null hypothesis would be true only 6.4% of the time.

ii. If this study were repeated many times and if the population means were all equal, we would see an *F* statistic as large as or larger than that observed in about 6.4% of the repetitions.

iii. If this study were repeated many times and if the population means were all equal, we would reject the null hypothesis in about 10% of the repetitions.

e. Suppose the necessary assumptions do hold. Using a 10% significance level, which of the following is the correct conclusion? Circle one:

[1]

i. It appears that the mean concentration levels are the same for the three city populations.

ii. It appears that at least one of the population mean concentration levels is different.

iii. It appears that the population mean concentration levels are all different.

f. Does there seem to be a difference between average sulfur dioxide concentration for City 1 versus City 2 at the 10% level?

Multiple Comparisons

Dependent Variable: concentration
Tukey HSD

(I) city	(J) city	Mean Difference (I-J)	Std. Error	90% Confidence Interval Lower Bound	Upper Bound
1	2	-4.0	1.3	-7.4	-.6
	3	-2.0	1.3	-5.4	1.4
2	1	4.0	1.3	.6	7.4
	3	2.0	1.3	-1.4	5.4
3	1	2.0	1.3	-1.4	5.4
	2	-2.0	1.3	-5.4	1.4

[2] Answer: ***YES*** ***NO***

Explain your answer (include relevant numbers from the output).

12. Response Rate – In a study, a random sample of 100 diabetic patients all received a new medical treatment for treating diabetes. At the end of the study, a total of 73 of the patients were reported as being a "responder" (that is, the patient responded favorably to the treatment).

a. The researcher had proposed the following hypotheses to be tested (before the data were collected): H_0: $p = 0.7$ versus H_a: $p > 0.7$.

[2] Clearly state what the symbol *p* in these hypotheses represents:

b. The researcher needs your help to conduct this test. Compute the corresponding test statistic and find the associated *p*-value based on the experiment results. Show your work.

[3]

Test statistic value: ________________

p-value: ________________

13. Name that Scenario: One important aspect in statistics is to understand which statistical methods or procedures are appropriate to use to address the research problem or question of interest. For each description of a research question below, select the letter corresponding to the statistical analysis technique most appropriate for addressing that research question. [12 points]

___ 1. A polling organization surveys 1000 people in a large community to assess if the average number of hours that people watch television is more than 5 hours a day. ___ 2. In another survey, the polling organization wants to determine if the percentage of men who identify themselves as socially conservative is higher than the percentage of women who identify as socially conservative. ___ 3. The polling organization surveys 1000 people in a large community to assess if the percentage of people in the community living in a mobile home is more than 10%. ___ 4. The polling organization surveys 1200 adults, records their age (in years) and how many hours they typically spend watching television per day, to assess if there is a significant relationship between age and hours of daily television viewing. ___ 5. The polling organization polls people on a regular basis to monitor their changing confidence in the economy. The organization creates a numerical index of confidence in the economy from answers to a questionnaire. Based on a random sample of people who were polled both in 2011 and again in 2012, the organization will assess if confidence in the economy changed from 2011 to 2012, on average. ___ 6. The polling organization conducts a survey of adults between the ages of 30 and 50 years old to assess if there is an association between marital status (married, divorced, never-married) and smoking status (smoker, non-smoker).	A. Simple linear regression B. 1-sample t-test for a population mean C. Paired t-test for a population mean difference D. 2-sample t-test for the comparison of two population means E. 1-sample Z-test for a population proportion F. 2-sample Z-test for the comparison of two population proportions G. Chi-squared test of goodness of fit H. Chi-squared test of independence I. Chi-squared test of homogeneity

Made in United States
Orlando, FL
04 November 2026

90203408R00114